Machine Guarding Handbook

A Practical Guide to OSHA Compliance and Injury Prevention

Frank R. Spellman
Nancy E. Whiting

Government Institutes
An imprint of
The Scarecrow Press, Inc.
Lanham, Maryland • Toronto • Oxford

Government Institutes

Published in the United States of America
by Government Institutes, an imprint of The Scarecrow Press, Inc.
A wholly owned subsidiary of
The Rowman & Littlefield Publishing Group, Inc.
4501 Forbes Boulevard, Suite 200
Lanham, Maryland 20706
www.govinstpress.com

Estover Road, Plymouth PL6 7PY, United Kingdom

British Library Cataloguing in Publication Information Available

Library of Congress Cataloging-in-Publication Data

ISBN 13: 978-0-86587-662-0 (pbk. : alk. paper)
ISBN 10: 0-86587-662-2 (pbk. : alk. paper)

™ The paper used in this publication meets the minimum requirements of American National Standard for Information Sciences—Permanence of Paper for Printed Library Materials, ANSI/NISO Z39.48-1992. Manufactured in the United States of America.

to

Robert E. Lee Spellman

and

Jefferson Davis Bates

Table of Contents

List of Figures and Tables

PREFACE

Many general safety and health books are written for supervisors and line managers, or for students making an in-depth study of safety and health. While these texts serve useful functions, they leave a large need unfilled—the need to provide a means for the safety practitioner or designated safety person to implement required OSHA safety compliance programs correctly. Tasked with implementing particular safety programs that require specific guidance, these people often need a basic overview of an applicable Standard's requirements to correctly implement all aspects of those safety programs in their facilities. *Machine Guarding Handbook* serves that purpose—a guide book to make compliance with machine guarding easier.

Machine guarding made easy? Why? What's the problem? Machine guarding seems like a reasonably straightforward task. Doesn't it?

Good questions.

The simple answer is machine guarding is an OSHA regulatory requirement—and nothing about following OSHA regulations is really altogether straightforward.

The compound answer is because the most common sources of mechanical hazards (and thus of injury) in the workplace are unprotected shafting, shaft ends, belt drives, gear train projections, rotating parts, chain and sprocket drives, and any exposed components of rotating parts—despite OSHA regulations and standards that cover machine guarding.

Remember that human body parts don't mix with machinery and moving parts—tearing, crushing, or scissor-like actions can affect anything (including body parts) that comes between the machine parts that perform those actions. Workers' risk of injury, of mutilation, of loss of limbs—or life—is very real.

Written in accessible plain English, *Machine Guarding Handbook* is designed to facilitate the effort required to incorporate a machine guarding safety program into the workplace, to meet OSHA requirements, and, most importantly, to protect workers on or around machinery.

Frank R. Spellman

Nancy E. Whiting

Acknowledgements

Special thanks to Joanne Drinan for her graphic support.

CHAPTER 1

INTRODUCTION

An Orrington, Maine company was hit with $20,500 in penalties from OSHA following the death of a maintenance worker in its trash-to-energy plant last year.

The worker was sweeping around an operating conveyor when the handle of his broom became caught in the nip point between a roller and the conveyor framework. The worker was pulled into the five-inch space created by the nip point and crushed. He died as a result of the accident.

An OSHA audit found that the conveyor wasn't properly equipped with machine guards that would have prevented the accident.

— Joy LePree, 1998

1.1 SETTING THE STAGE

When 29 CFR 1910 Subpart O of the General Industry Standards describing machine safeguarding requirements came out in November 1975 from the federal Occupational Safety and Health Administration (OSHA), all American employers were expected to comply with those provisions. But American workplaces continue to be plagued with an unacceptable frequency and severity of injuries, often resulting in amputation (or worse), from contact with industrial machines. After more than 20 years of federal, state, and local inspections, we see less than satisfactory results. In fact, the National Safety Council's publication, Safety and Health gives the U.S. a C+ grade in workplace safety (p 37).

Machines are essential. With machines, workers accomplish tasks unheard of before their invention. While machines do allow for more efficient, complex, and productive work, they must be used with a certain degree of caution (some machine use requires a ***great*** deal of caution).

Witnessing an on-the-job accident involving machine damage to workers might include seeing—and providing first aid for—crushed fingers or hands, severed fingers, burns, scalping, blindness, arm or leg amputations—or worse (see Figure 1.1). We've said enough about the bloody, gory details, but remember this—the list of possible machinery-related injuries is as long as it is horrifying. More than 50% of all machine-related injuries result in permanent partial disability, and very often, in amputation of some part of the body (Brubaker, 1997).

Typically, whenever an accident occurs in the workplace, the organization's safety professional or designated safety official is called upon to investigate the incident. Even the most experienced, seasoned, hardened, tough-as-nails safety person hates to investigate injuries involving machines. Why? Because from experience they know two things:

- such incidents are almost always bloody, gory, nasty—sometimes beyond description, and
- they also know that such incidents should never occur in the first place.

They shouldn't occur . . . but they do. The fact is, they occur far too often.

Safety professionals in the field know that accidents involving machines occur. Why? Good question. We know that without proper machine guarding around equipment such as conveyors, mechanical and hydraulic stamping presses, press brakes, industrial robots, molding presses, packaging equipment, welding, riveting, forming, and other automated production applications, workers can easily become caught in machinery or struck by flying sparks and debris. Machine guards prevent—or help prevent—such incidents.

Whatever the type of machinery involved, the basic hazards fall into categories. Machine components that move rapidly with power, or points of operation where the machine performs its work, are typical hazard sources. According to Ted Ferry (1990) the most common sources of mechanical hazards are unguarded shafting, shaft ends, belt drives, gear trains, and/or projections on rotating parts. Locations where chain and moving parts pass a stationary part or another moving part can create a scissor-like effect on anything caught between the parts.

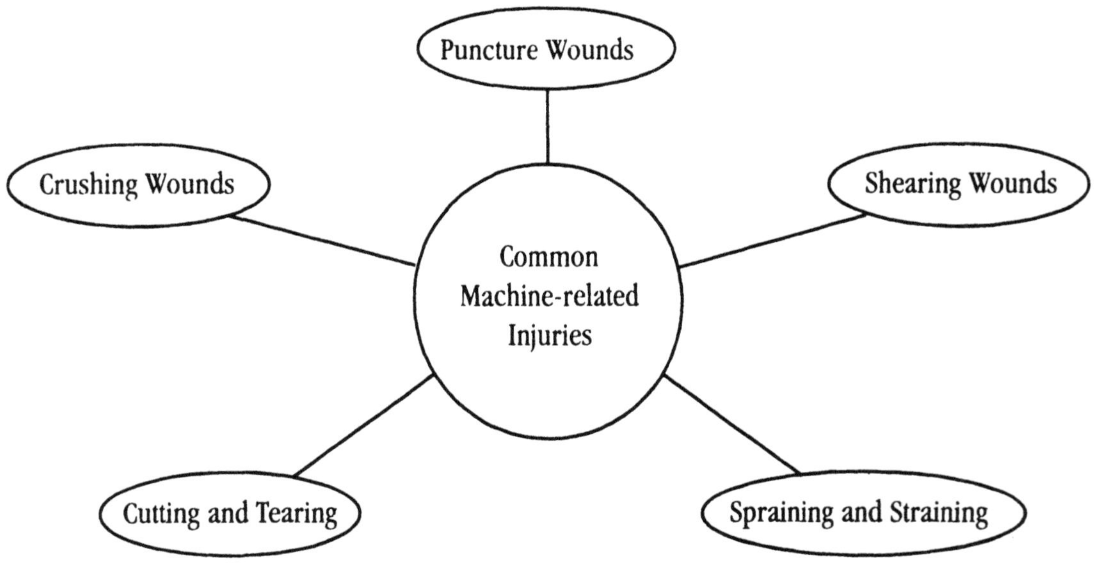

Figure 1.1. Some common machine-related injuries.

Many people commonly associate machinery and machine guarding only with light or heavy industrial operations. We point out, however, that factories or other industrial-type worksites are not the only places where machines are used, or where machines cause injuries. Machines are used in many, many locations, and wherever they are used, injuries from them can occur.

One industry where machine-related incidents are quite common (and frequently devastating) is in agriculture. With even minimal knowledge about modern agricultural techniques, we have little difficulty associating farms with powerful farm equipment (tractors, mowers, shredders, harvesters, grinders, blowers, augers, and balers). These machines create many possible hazards for both operators and bystanders. For example, agricultural workers can be cut, crushed, pulled into, or struck by objects thrown from these machines. Farm equipment has cutting edges, gears, chains, revolving shafts, rotating blades, levers, and similar hazards. Many injuries occur when farmers fall while working from or near these machines. Often the effects of less serious injuries are compounded by the time help finally arrives to an injured farmer working alone in a remote or isolated location.

> *Note: According to the National Institute for Occupational Safety and Health, each year, 100,000 children and youths are injured on farms—and more than 100 are killed. The National Committee for Childhood Agricultural Injury Prevention reports the primary causes of fatal and nonfatal injuries to children on farms are (listed in order of highest frequency of occurrence): farm tractors, livestock, farm machinery, building structures, and falls.*

A number of farmers lose fingers, hands, arms, and even feet in farm equipment accidents each year. Some lose their lives. Nearly all of the accidents are preventable, but risk-taking, carelessness, or lack of knowledge allow them to occur. Agriculture isn't an exception. No matter what the industry, such tragedies can be prevented, especially if workers understand the hazards machines present, and practice the safety precautions needed to avoid them.

Even though manufacturers take many steps to make machinery safe, removing all hazards is simply not possible. Some machine parts cannot be completely shielded and still perform their function. For instance, a totally enclosed cutting blade could not cut. Guards must be removed for maintenance, but for a variety of reasons, they sometimes don't get replaced.

One of the key elements that help to prevent the occurrence of machine-type injuries is proper safeguarding of such equipment. Simply stated, safeguards are essential for protecting workers from these needless and preventable injuries.

Earlier we stated that injuries from using or maintaining machines occur far too often. We also point out that machine guarding and related violations continuously rank among the top 10 OSHA citations issued. In fact, 29 CFR 1910.219 "Mechanical Power Transmission" and 29 CFR 1910.212 "Machine Guarding: General Requirements" were the number six (6) and number seven (7) top OSHA violations for fiscal year 1997, with more than 3,000 federal citations issued for each. Note that mechanical power presses have also become an area of increasing concern. In April 1997, OSHA launched a National Emphasis Program on Mechanical Power Presses (CPL 2-1.24). This program targets industries that have high amputation rates, and includes both education and enforcement efforts.

In *Machine Guarding Handbook*, we familiarize you with the hazards of unguarded machines, common safeguarding methods, the safeguarding of machines and robots, why guarding is so important, the varying methods of machine guarding, and how to implement safeguarding techniques. We also provide important regulatory information, including specific information on mechanical power presses and guidelines for robotics safety (because of their proliferation in usage, the safe operation and use of robots and robotic systems is an area of increasing concern for both the safety professional and OSHA compliance officer).

Of special importance are the standard checklist items that should be used by the organizational safety person or other designated official to ensure full compliance with safeguarding requirements. You can be certain that the OSHA compliance officer will be using a similar checklist to ensure that you are indeed in compliance with machine guarding standards. We also provide a sample lockout/tagout program.

If we had to sum up the exact purpose of machine guarding, we could do so by stating, "Preventing contact of the human body with dangerous parts of machines is the basic purpose of machine guarding. When fingers, arms, hair, or other body parts enter into or make contact with moving machinery, the results can be ugly, disastrous—and sometimes fatal."

From our experience with safety audits, we find that when machines are missing guards, or when guards are not provided for dangerous machines, the common replies from the persons in charge are:

- "machine guards are too expensive"
- "if workers are careful, they will not be injured"
- "machine guards interfere with the work"
- "oh, well, nothing more than common sense is required to prevent the machine from injuring my workers."

Our common sense tells us that anyone who can use these excuses (and they are excuses, not reasons) hasn't thought about the value of what machine safeguards are protecting. Asked yourself: "How expensive is a worker's arm, leg, fingers, scalp, eyes—their life—in comparison to a machine guard?" We learned the answer to that question long ago—the first time we saw a worker who had lost part of a finger in an unguarded machine.

The theme of *Machine Guarding Handbook* is that machine guards are essential for protecting workers from needless and preventable injuries. The purpose of *Machine Guarding Handbook*: to provide the user of this text with a straightforward, plain English presentation of the material required to implement a viable machine guarding program within your organization (Figure 1.2).

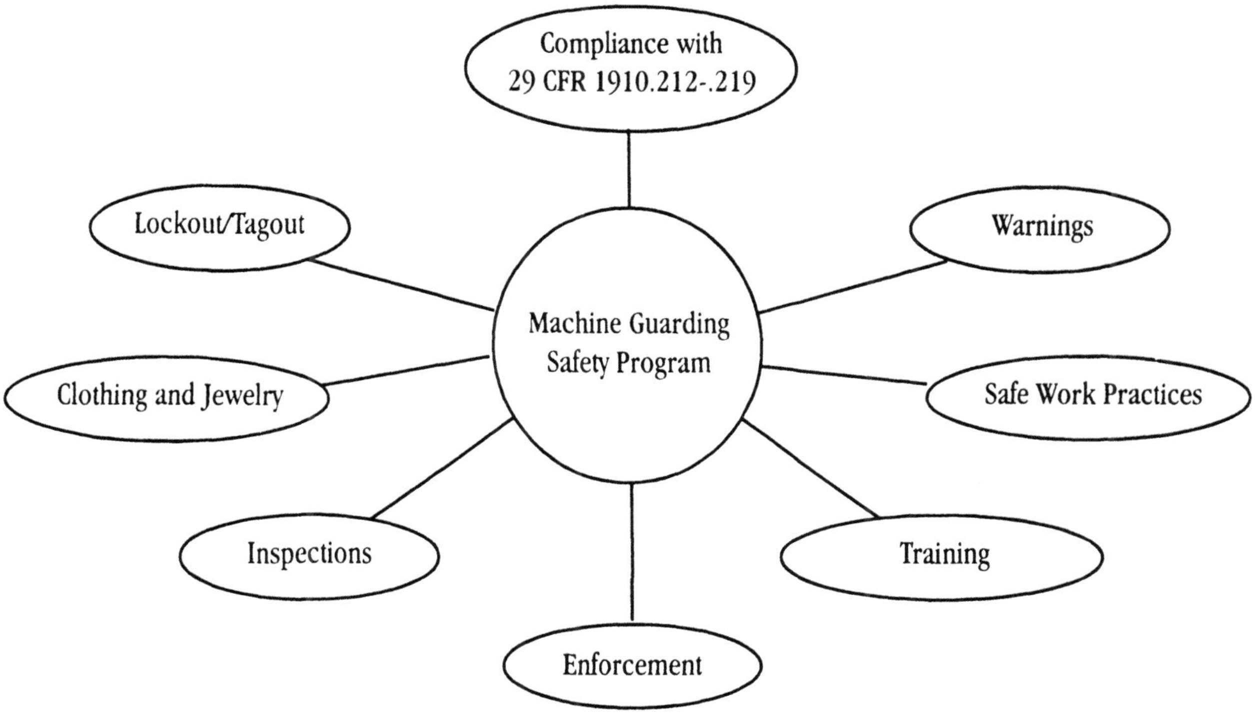

Figure 1.2. Elements of Machine Guarding Safety Program.

1.2 SCOPE OF TEXT

Spanning coverage of safeguarding protection devices/methods from the simplest machine to the most sophisticated robotic system, *Machine Guarding Handbook* makes a strong argument that safety must be engineered into mechanized systems, and tailored to specific technologies and processes. *Machine Guarding Handbook* presents both an overview and a very practical guide, featuring a wide range of techniques important to the organizational safety official responsible for safety compliance in manufacturing and process industries. It offers a wealth of guidelines, checklists, and safety data crucial for safety officials and technicians who want to institute a well-planned, organized, systematic, and meticulously executed machine-guarding program.

> *Note: Much of the information in this text has been adapted from* Concepts and Techniques of Machine Safeguarding, *published by the Occupational Safety and Health Administration (publication no. 3067, revised 1992). This excellent reference source is available from the Superintendent of Documents, U.S. Government Printing Office, Washington, D.C. 20402.*

1.3 GENERAL SAFEGUARDING REQUIREMENTS FOR ALL MACHINES

The operation of machinery is serious business, and should be treated seriously. Avoiding any type of machinery-related injury requires that strict safety practices are employed—safety practices in line with OSHA requirements.

OSHA states that at least one method of machine guarding must be provided to protect workers in the machine areas from hazards created by point of operation, ingoing nip points, rotating parts, and flying chips and sparks. Examples of guarding methods include barrier guards, two-hand tripping devices, safety devices for robotics, electronic safety devices, etc.

Whatever type of guard is used, it must be affixed to the machine where and when possible, or secured elsewhere if for any reason attachment to the machine is not possible. The guard itself must be constructed, configured, and/or placed so that it does not present an accident hazard itself.

Guarding must be provided at the point of operation—the area on a machine where work is actually performed upon the material processed. The guarding device must conform to any appropriate standards, or in the absence of applicable specific standards, must be designed and constructed to prevent operators from having any part of their bodies in the danger zone during the operating cycle. When special hand tools are used for placing and removing material, the tools must be constructed so they do not interfere with handling of materials, and they must prevent operators from placing any part of their body in the danger zone. These tools cannot replace other required guarding, only supplement the protection provided.

Guarded machinery may be roughly categorized as:

- Forming, punching, and shearing machines, including power and hand presses, forging hammers, power brakes, and shears.
- Milling, drilling, and grinding machines, including drill presses, boring mills, lathes, shakers, and grinders.
- Sawing, ripping, and planing machines, including circular and band saws, jointers, and planers.

Other general machine guarding requirements include:

- Revolving barrels, containers, and drums must be guarded by an enclosure that is interlocked with the drive mechanism, so that the barrel, drum, or container cannot revolve unless the guard enclosure is in place.
- When the periphery of fanblades is less than seven (7) feet above the floor or working level, the blades must be guarded. The guard openings must not be larger than one-half (1/2) inch.
- All machines designed for a fixed location must be securely anchored to prevent 'walking' or moving.

1.4 MACHINE-GUARDING DEFINITIONS

In any specialized field, learning the vocabulary is essential for understanding. Machine guarding is no different. In this section, we define many of the terms most commonly used and/or associated with machine guarding in general, and those terms used specifically in *Machine Guarding Handbook*. For those requiring a complete listing of terms and definitions, refer to 29 CFR 1910.211.

Point of operation: the area on the machine where work is actually performed upon the material being processed. This area requires guarding if an injury to an employee is possible. These devices should be designed and constructed to prevent operators from having any part of their bodies in the danger zone during the operating cycle. Machines requiring point-of-operation protection are guillotine cutters, shears, power presses, milling machines, power saws, jointers, portable power tools, forming rolls, and calendars.

Shear Points: created when the edges of two objects are moved closely enough together to cut a soft material, as in the case of a pair of shears or an auger. Cutting points are created when a single object moves forcefully or rapidly enough to cut, as in the case of a sickle blade. Both shear and cutting points are created on machinery designed to cut, as in harvesters, and on those that are not designed to cut, as in an auger. These points present hazards because of their cutting force, and because they often move so rapidly that they may not be visible. Forgetting that the machine is operating is easy.

Note: Because some cutting and shearing points cannot be guarded, operators must remain aware of their hazard, and be especially alert when they are operating such machines. Operators should also know to warn others unfamiliar with the machine and to look out for their safety, because of the danger of thrown objects while using cutting-type equipment.

Pinch Point: formed when two rotating objects move together, and at least one of them moves in a circle. For example, the point at which a belt runs onto a pulley is a pinch point. Belt drives, chain drives, and gear drives are other sources of pinch points in power transmission devices. Feed rolls, gathering chains, and similar equipment to draw crops into a machine also create pinch points. Fingers, hands, and feet can be caught directly in pinch points, or they may be drawn into the pinch points by loose clothing that becomes entangled. Contact may be made by simply brushing against unshielded parts or by falling against them. Workers can become entangled in pinch points if they take chances and reach or work near rotating parts. Machines move too fast and powerfully for the operator to get out of a pinch point once caught in it.

Note: To avoid injury from pinch points, be aware of the areas where pinch points occur and avoid them. Wear clothing that fits well and is not loose or floppy. Never reach over or work near rotating parts. Turn off machinery to work on it. Always replace shields if you must remove them for maintenance.

Case Study 1.1

Though old, the printing press at a small print shop in Lancaster, PA is still a powerful machine. An experienced pressman in the well-trained workforce lost part of a finger in it in 1996—because of a change in gloves.

The pressmen wore surgical-type gloves for many years, and were accustomed to handling the inks and paper with them. When they switched to thicker gloves to prevent ink leakage, they were all aware of the different handling characteristics the gloves created. The surgical gloves fit skin-tight. The new gloves did not. But months went by with no incident connected with the gloves, and the pressmen became accustomed to them. Work again became automatic.

Shortly after the start of the workday on January 15, 1996, the tips of the three last fingers of the glove on a pressman's right hand caught in the rollers of the feed, and pulled his hand into the press. He was lucky. He pulled back instantly and while the glove lost most of three fingers, he only lost down to the first knuckle of his right little finger.

> **Wrap points**: rotating shafts are the most common source of wrap point accidents, although any exposed machine part that rotates can be a wrap point. A cuff, sleeve, pant leg or even just a thread can catch on a rotating part and result in serious injury. Entanglement with a wrap point can pull a worker into the machine, or clothing may become so tightly wrapped that workers are crushed or suffocated. Workers can be thrown off balance and fall into other machine parts. Even a perfectly round shaft can be a hazard with enough pressure to hold clothing against the shaft. Shafts that are not round increase the hazard significantly. Clothing is more likely to catch if the shaft is not clean—a little mud or dried manure, or a nick on the shaft can provides a snag-point for textiles to momentarily catch and become entangled. Ends of shafts which protrude beyond bearings are also dangerous. Universal joints, keys, and fastening devices can also snag clothing.

Such accidents are not restricted to industry. Isadora Duncan, the creator of modern dance, was killed while a passenger in an open car. When the end of the long scarf she was wearing caught in the rear axle of the Bugatti racer and entangled, it swiftly snapped her neck. She died instantly.

> *Note: All equipment should be checked for wrap points; those that can be shielded should be. Warnings should be placed on those that can't be covered, or painted a bright color or adorned with warning stripes. All workers should be made aware of wrap points and be alert to their danger.*

> **Crush points**: created when two objects move toward each other, or when one object moves toward a stationary one. For example, in agricultural work, hitching a tractor to an implement creates a potential crush point.

Failure to block up equipment safely can result in a fatal crushing injury. A jack may slip, a hose or overhead support may break, or the equipment may roll. Workers must ensure they take extra precautions when working with raised machinery.

The head or chest of an operator may be crushed between moving equipment and a low beam or other part of a building.

> *Note: To prevent being crushed or pinned, first, recognize the potentially dangerous situations. Then, avoid them whenever possible. Block all machinery securely if it must be worked under. If an implement can roll freely, block its wheels so it cannot roll.*

Free-Wheeling Parts: many machine parts continue to spin after the power is shut off; for instance, flywheels, fans, rotary mower blades, hammer mills of feed grinders, and cutter heads of forage harvesters. Never touch these parts until they have stopped moving completely. This may take as long as two or three minutes.

Springs: under compression, springs will expand with great force when released, and stretched springs will contract rapidly when released. Know what direction a spring will move, and how it might affect another machine part when released—and stay out of its path.

Hydraulic systems: contain fluid under very high pressure. Before loosening, tightening, removing, or otherwise working with any fittings or parts, relieve this pressure. Jet streams from even pinhole leaks can penetrate flesh. In addition, the liquid is often hot. Before attempting any service on hydraulic systems, shut off the engine that powers the hydraulic pump. Lower the implement to the ground and relieve the pressure. Follow the instructions in the operator's manual, because following specific procedures for servicing the systems are very important to workers' safety.

> *Note: Unless the context clearly requires otherwise, the following woodworking machinery terms have the meaning prescribed (29 CFR 1910.215).*

Push stick: a narrow strip of wood or other soft material with a notch cut into one end that is used to push short pieces of material through saws.

Block: a short block of wood, provided with a handle similar to that of a plane and with a shoulder at the rear end, used for pushing short stock over revolving cutters.

Flanges: collars, discs, or plates between which wheels are mounted. They are referred to as adaptor, sleeve, or back up type.

Snagging: grinding which removes relatively large amounts of material without regard to close tolerances or surface finish requirements.

Off-hand grinding: the grinding of any material or part which is held in the operator's hand.

Safety guard: an enclosure designed to restrain the pieces of the grinding wheel, furnishing all possible protection in the event that the wheel is broken in operation.

Type 1 straight wheels: wheels having diameter, thickness, and hole size dimensions. They should be used only on the periphery. Note: Type 1 wheels must be mounted between flanges.

Cutting off wheels: wheels having diameter thickness and hole size dimensions. These are subject to all mounting and use limitations listed for Type 1 wheels. They may be steel centered, diamond abrasive, or organic bonded abrasive of the plain or reinforced type.

Limitation: Cutting off wheels are recommended only for use on specially designed and fully guarded machines, and are subject to the following maximum thickness and hole size limitations.

Wheel Diameter	Maximum Thickness (inches)
< 6 inches	3/8
6 to 12 inches	1/4
12 to 23 inches	3/8
> 23 inches	1/2

Maximum hole size for cutting-off wheels should not be larger than 1/4-wheel diameter.

Abrasive wheel: a cutting tool consisting of abrasive grains held together by organic or inorganic bonds. Diamond and reinforced wheels are included.

Note: Immediately before an abrasive wheel is mounted to any machine normally equipped with such devices, it should be thoroughly inspected and sounded by the user (ring test) to make sure no damage occurred in transit, storage, or otherwise.

Ring test: to sound an abrasive wheel before mounting. Before sounding an abrasive wheel, ensure the wheel is dry and free from sawdust, otherwise the sound will be deadened. Note that organic bonded wheels do not emit the same clear metallic ring as do vitrified and silicate wheels.

In conducting the ring test, tap the wheel about 45° on each side of the vertical centerline and about 1 or 2 inches from the periphery as indicated in Figure 1.3.

Figure 1.3. Standard grinding wheel divided into 4 sections, ready for ring testing.

Then rotate the wheel 45° and repeat the test. A sound and undamaged wheel will give a clear metallic tone. If cracked, the wheel will emit a dead sound, not a clear "ring."

Note: Workers must know how to properly conduct an abrasive wheel "ring test." OSHA auditors commonly interview workers and ask them to demonstrate how to perform the ring test.

Organic wheels: wheels which are honed by means of an organic material such as resin, rubber, shellac, or other similar bonding agent.

Inorganic wheels: wheels which are honed by means of inorganic material such as clay, glass, porcelain, or related ceramic materials are characterized as **vitrified bonded wheels.**

Note: Unless the context clearly requires otherwise, the following terms and definitions refer to mills and calendars in the rubber and plastic industries (29 CFR 1910.216).

Bite: the nip point between any two in-running rolls.

Calendar: a machine equipped with two or more metal rolls revolving in opposite directions, used for continuously sheeting or plying up rubber and plastics compounds and for frictioning or coating materials with rubber and plastics compounds.

Mill: a machine consisting of two adjacent metal rolls set horizontally, which revolve in opposite directions (i.e., toward each other as viewed from above) used for the mechanical working of rubber and plastics compounds.

Note: Unless the context clearly requires otherwise, the following terms apply to power presses (29 CFR 1910.217).

Antirepeat: the part of the clutch/brake control system designed to limit the press to a single stroke if the tripping means is held operated. Antirepeat requires release of all tripping mechanisms before another stroke can be initiated. Antirepeat is also called single stroke reset or reset circuit.

Brake: the mechanism used on a mechanical power press to stop and/or hold the crankshaft, either directly or through a gear train, when the clutch is disengaged.

Bolster plate: the plate attached to the top of the press having drilled holes or T-slots for attaching the lower die or die shoe.

Clutch: the coupling mechanism used on a mechanical power press to couple the flywheel to the crankshaft, either directly or through a gear train.

Full revolution clutch: a type of clutch that when tripped, cannot be disengaged until the crankshaft has completed a full revolution and the press slid a full stroke.

Part revolution clutch: a type of clutch that can be disengaged at any point before the crankshaft has completed a full revolution and the press slid a full stroke.

Direct drive: the type of driving arrangement wherein no clutch is used; coupling and decoupling of the driving torque is accomplished by energization and de-energization of a motor. Even if a clutch is not emplyed, direct drives match the operational characteristics of "part revolution clutches" because the driving power may be disengaged during the stroke of the press.

Concurrent: acting in conjunction: used to describe a situation wherein two or more controls exist in an operated condition at the same time.

Continuous: uninterrupted multiple strokes of the slide without intervening stops (or other clutch control action) at the end of individual strokes.

Counterbalance: the mechanism used to balance or support the weight of the connecting rods, slide, and slide attachments.

Device: a press control or attachment that:

- restrains the operator from inadvertently reaching into the point of operation
- prevents normal press operation if the operator's hands are inadvertently within the point of operation
- automatically withdraws the operator's hands if the operator's hands are inadvertently within the point of operation as the dies close
- prevents the initiation of a stroke, or stops a stroke in progress, when an intrusion through the sensing field occurs from any part of the operator's body or by any other object.

Presence sensing device: a device designed, constructed, and arranged to create a sensing field or area that signals the clutch/brake control to deactivate the clutch and activate the brake or the press when any part of the operator's body or a hand tool is within such a field or area.

Gate or movable barrier device: a movable barrier arranged to enclose the point of operation before the press stroke can be started.

Holdout or restraint device: a mechanism, including attachments for operator's hands, that when anchored and adjusted prevents the operator's hands from entering the point of operation.

Pull-out device: a mechanism attached to the operator's hands, that when properly adjusted and connected to the upper die or slide of the press, is designed to withdraw the operator's hands as the dies close, if the operator's hands are inadvertently within the point of operation.

Sweep device: a single or double arm (rod) attached to the upper die or slide of the press, designed to move the operator's hands to a safe position as the dies close, if the operator's hands are inadvertently within the point of operation.

Two-hand control device: a two-hand trip that further requires concurrent pressure from both hands of the operator during a substantial part of the die-closing portion of the stroke and the press.

Die: the tooling used in a press for cutting or forming material. An upper and a lower die make a complete set.

Die builder: any person who builds dies for power presses.

Die set: a tool holder held in alignment by guide posts and bushings, consisting of a lower shoe, an upper shoe or punch holder, and the guide posts and bushings.

Die setter: an individual who places dies in, or removes dies in or from, mechanical power presses, and who, as a part of his duties, makes the necessary adjustments to cause the tooling to function properly and safely.

Die setting: the process of placing dies in, or removing dies in or from, a mechanical power press, and the process of adjusting the dies or other tooling and safeguarding means to cause them to function properly and safely.

Die shoe: a plate or block upon which a die holder is mounted. A die shoe functions primarily as a base for the complete die assembly, and when used, is bolted or clamped to the bolster plate or the face of the slide.

Ejector: a mechanism for removing work or material from between the dies.

Face of slide: the bottom surface of the slide to which the punch or upper die is generally attached.

Feeding: the process of placing material in, or removing material from the point of operation.

Automatic feeding: feeding wherein the material or part being processed is placed within or removed from the point of operation by a means not requiring action by an operator on each stroke of the press.

Semiautomatic feeding: feeding wherein the material or part being processed is placed within or removed from the point of operation by an auxiliary means controlled by an operator on each stroke of the press.

Manual feeding: feeding wherein the material or part being processed is handled by the operator on each stroke of the press.

Foot control: a foot-operated control mechanism designed to be used with a clutch or clutch/brake control system.

Foot pedal: a foot-operated lever designed to operate the mechanical linkage that trips a full revolution clutch.

Guard: a barrier that prevents entry of the operator's hands or fingers into the point of operation.

Die enclosure guard: an enclosure attached to the die shoe or stripper, or both, in a fixed position.

Fixed barrier guard: a die space barrier attached to the press frame.

Interlocked press barrier guard: a barrier attached to the press frame, interlocked so that the press stroke cannot be started normally unless the guard itself (or its hinged or movable sections) enclose the point of operation.

Adjustable barrier guard: a barrier requiring adjustment for each job or die setup.

Guide post: the pin attached to the upper or lower die shoe operating within the bushing on the opposing die shoe, to maintain the alignment of the upper and lower dies.

Hand feeding tool: any hand-held tool designed for placing material or parts in, or removing material or parts from the point of operation.

Inch: an intermittent motion imparted to the slide (on machines using part revolution clutches) by momentary operation of the **inch operating means.** Operation of the inch operating means consists of engaging the driving clutch so that a small portion of one stroke or indefinite stroking can occur, depending upon the length of time the inch operating means is operated. **Inch** is a function used by the die setter to setup dies and tooling, but is not intended for use during production operations by the operator.

Jog: an intermittent motion imparted to the slide by momentary operation of the drive motor, after the clutch is engaged with the flywheel at rest.

Knockout: a mechanism for releasing material from either die.

Liftout: the mechanism also known as knockout.

Operator's station: the complete complement of controls used by or available to an operator on a given operation for stroking the press.

Pinch point: any point other than the point of operation at which a part of the body could be caught between the moving parts of a press or auxiliary equipment, or between moving and stationary parts of a press or auxiliary equipment, or between the material and moving part or parts of the press or auxiliary equipment.

Point of operation: the area of the press where material is actually portioned and work is being performed during any process such as shearing, punching, forming, or assembling.

Press: a mechanically powered machine that shears, punches, forms or assembles metal or other material by means of cutting, shaping, or combination dies attached to slides. A press consists of a stationary bed or anvil, and a slide (or slides) having a controlled reciprocating motion toward and away from the bed surface, the slide being guided in a definite path by the frame of the press.

Repeat: an unintended or unexpected successive stroke of the press resulting from a malfunction.

Safety block: a prop that, when inserted between the upper and lower dies or between the bolster plate and the face of the slide, prevents the slide from falling of its own dead-weight.

Single stroke: one complete stroke of the slide, usually initiated from a full open (or up) position, followed by closing (or down), and then a return to the full open position.

Single stroke mechanism: an arrangement used on a full revolution clutch to limit the travel of the slide to one complete stroke at each engagement of the clutch.

Slide: the main reciprocating press member. A slide is also called a ram, plunger, or platen.

Stop control: an operator control designed to immediately deactivate the clutch control and activate the brake to stop slide motion.

Stripper: a mechanism or die part for removing the parts or material from the punch.

Stroking selector: the part of the clutch/brake control that determines the type of stroking when the operating means is actuated. The stroking selector generally includes positions for "Off" (Clutch Control), "Inch," "Single Stroke," and "Continuous" (when Continuous is furnished).

Trip or (tripping): activation of the clutch to "run" the press.

Turnover bar: a bar used in die-setting to manually turn the crankshaft of the press.

Two-hand trip: a clutch-actuating means requiring the concurrent use of both hands of the operator to trip the press.

Utilized tooling: a type of die in which the upper and lower members are incorporated into a self-contained unit that holds the die members in alignment.

Control system: includes sensors, manual input and mode selection elements, interlocking and decision-making circuitry, and output elements to the press operating mechanism.

Brake monitor: a sensor designed, constructed, and arranged to monitor the effectiveness of the press braking system.

Presence sensing device initiation: an operating mode of indirect manual initiation of a single stroke by a presence sensing device. The device senses that work motions of the operator related to feeding and/or removing parts are completed, and all parts of the operator's body or hand tools are safely clear of the point of operation.

Safety system: the integrated total system: Includes the pertinent elements of the press, the controls, the safeguarding, and any required supplemental safeguarding and their interfaces with the operator as well as the environment. Designed, constructed, and arranged to operate together as a unit, such that a single failure or single operating error will not cause injury to personnel due to point of operation hazards.

Authorized person: one to whom the authority and responsibility to perform a specific assignment has been given by the employer.

Certification or **certify**: in the case of design certification/validation, that the manufacturer has reviewed and tested the design and manufacture, and in the case of installation certification/validation and annual recertification/revalidation, that the employer has reviewed and tested the installation, and concludes in both cases that the requirements of 29 CFR 1910.2.17 (a) through (h) and Appendix A have been met. The certifications are made to the validation organization.

Certification/validation and **certify/validate**: the combined process of certification and validation.

Note: Unless the context clearly requires otherwise, the following terms apply to forging and hot metal applications.

Forging: the product of work on metal formed to a desired shape by impact or pressure in hammers, forging machines (upsetters), presses, rolls, and related forming equipment. Forging hammers, counterblow equipment and high-energy-rate forging machines impact the workpiece. While most other types of forging equipment and high-energy-rate forging machines impact the workpiece, other types of forging equipment also impart squeeze pressure in shaping the stock. Some metals can be forged at room temperature, but the majority of metals are made more plastic for forging by heating.

Open framehammers (or blacksmith hammers): hammers used primarily for the shaping of forgings by impact with flat dies. Open frame hammers are generally so constructed that the anvil assembly is separate from the operating mechanism and machine supports; it rests on its own independent foundation. Certain exceptions are forging hammers made with the frame mounted on the anvil (e.g., the smaller, single-frame hammers are usually made with the anvil and frame in one piece).

Steam hammers: a type of drop hammer where the ram is raised for each stroke by a double-action steam cylinder and the energy delivered to the workpiece is supplied by the velocity and weight of the ram and attached upper die driven downward by steam pressure. Energy delivered during each stroke may be varied.

Gravity hammers: a class of forging hammer wherein energy for forging is obtained by the mass and velocity of a freely-falling ram and the attached upper die. Examples: board hammers and air-lift hammers.

Forging presses: a class of forging equipment wherein the shaping of metal between dies is performed by mechanical or hydraulic pressure, and usually is accomplished with a single workstroke of the press for each die station.

Trimming presses: a class of auxiliary forging equipment that removes flash or excess metal from a forging. This trimming operation can also be done cold, as can coining, a product-sizing operation.

High-energy-rate forging machines: a class of forging equipment wherein high ram velocities resulting from the sudden release of a compressed gas against a free piston impart impact to the workpiece.

Forging rolls: a class of auxiliary forging equipment wherein stock is shaped between power-driven rolls bearing contoured dies. Usually used for preforming, roll forging is often employed to reduce thickness and increase length of stock.

Ring rolls: a class of forging equipment used for shaping weldless rings from pieced discs, or thick-walled, ring-shaped blanks between rolls that control wall thickness, ring diameter, height, and contour.

Bolt-headers: the same as an upsetter on forging machines, except that the diameter of stock fed into the machine is much smaller (commonly 3/4 inch or less).

Rivet making machines: the same as upsetters and boltheaders when producing rivets with stock diameter of 1-inch or more. Rivet-making with less than 1-inch diameter is usually a cold forging operation.

Upsetters (or forging machines, or headers): a type of forging equipment related to the mechanical press, in which the main forming energy is applied horizontally to the workpiece which is gripped and held by prior action of the dies.

> *Note: Unless the context clearly requires otherwise, the following terms apply to mechanical power-transmission guarding applications.*

Belts: include all power transmission belts, such as flat belts, round belts, V-belts, etc., unless otherwise specified.

Belt shifter: a device for mechanically shifting belts from tight to loose pulleys or vice versa, or for shifting belts on cones of speed pulleys.

Belt pole (sometimes called a **belt shipper** or **shipper pole**): a device used in shifting belts on and off fixed pulleys on line, or countershafts where there are no loose pulleys.

Exposed to contact: the location of an object in such a location that a person is likely to come into contact with it and be injured.

Flywheels: include balance wheels, and flywheel pulleys mounted and revolving on crankshaft or engine, or other shafting.

Maintenance runway: any permanent runway or platform used for oiling, maintenance, running adjustment, or repair work, but not for passageway.

Nip-point belt and pulley guard: a device that encloses the pulley and is provided with rounded or rolled edge slots through which the belt passes.

Point of operation: the point at which cutting, shaping, or forming is accomplished upon the stock, and including such other points as may offer a hazard to the operator in inserting or manipulating the stock in the operation of the machine.

Prime movers: include steam, gas, oil, and air engines, motors, steam, and hydraulic turbines, and other equipment used as a source of power.

Sheaves: grooved pulleys, so classified unless used as flywheels.

Note: The following terms apply to robots and robotics.

Actuator: a power mechanism used to effect motion of the robot; a device that converts electrical, hydraulic, or pneumatic energy into robot motion.

Application program: the set of instructions that defines the specific intended tasks of robots and robot systems. This program may be originated and modified by the robot user.

Attended continuous operation: the time when robots are performing (production) tasks at a speed no greater than slow speed through attended program execution.

Attended program verification: the time when a person within the restricted envelope (space) verifies the robot's programmed tasks at programmed speed.

Automatic mode: the robot state in which automatic operation can be initiated.

Automatic operation: the time during which robots are performing programmed tasks through unattended program execution.

Awareness barrier: physical and/or visual means that warns a person of a present or approaching hazard.

Awareness signal: a device that warns a person of a present or approaching hazard by means of an audible sound or visible light.

Axis: the point about which a rotating body such as a tool turns.

Barrier: a physical means of separating persons from the restricted envelope (space).

Control device: any piece of control hardware providing a means for human intervention in the control of a robot or robot system, such as an emergency-stop button, a start button, or a selector switch.

Control program: the inherent set of control instructions that defines the capabilities, actions, and responses of the robot system. This program is usually not intended to be modified by the user.

Coordinated straight line motion: control wherein the axes of the robot arrive at their respective end points simultaneously, giving a smooth appearance to the motion. Control wherein the motions of the axes are such that the Tool Center Point (TCP) moves along a prespecified type of path (line, circle, etc.).

Device any piece of control hardware such as an emergency-stop button, selector switch, control pendant, relay, solenoid valve, sensor, etc.

Drive power: the energy source or sources for the robot actuators.

Emergency stop: the operation of a circuit using hardware-based components that overrides all other robot controls, removes drive power from the robot actuators, and causes all moving parts to stop.

Enabling device: a manually operated device that permits motion when continuously activated. Releasing the device stops robot motion and motion-associated equipment that may present a hazard.

End-effector: an accessory device or tool specifically designed for attachment to the robot wrist or tool mounting plate to enable the robot to perform its intended task. (Examples may include gripper, spot-weld gun, arc-weld gun, spray-paint gun, or any other application tools).

Energy source: any electrical, mechanical, hydraulic, pneumatic, chemical, thermal, or other source.

Envelope (Space), Maximum: the volume of space encompassing the maximum designed movements of all robot parts, including the end-effector, workpiece, and attachments.

Restricted envelope (space): that portion of the maximum envelope to which a robot is restricted by limiting devices. The maximum distance that the robot can travel after the limiting device is actuated defines the boundaries of the restricted envelope (space) of the robot.

Note: The safeguarding interlocking logic and robot program may redefine the restricted envelope (space) as the robot performs its application program.[1]

Operating envelope (space): that portion of the restricted envelope (space) that is actually used by the robot while performing its programmed motions.

Hazard: a situation that is likely to cause physical harm.

Hazardous motion: any motion that is likely to cause personal physical harm.

Industrial equipment: physical apparatus used to perform industrial tasks, such as welders, conveyors, machine tools, fork trucks, turn tables, positioning tables, or robots.

Industrial robot: a reprogrammable, multifunctional manipulator designed to move material, parts, tools, or specialized devices through variable programmed motions for the performance of a variety of tasks.

Industrial robot system: a system that includes industrial robots, the end-effectors, and the devices and sensors required for the robots to be taught or programmed, or for the robots to perform the intended automatic operations, as well as the communication interfaces required for interlocking, sequencing, or monitoring the robots.

Interlock: an arrangement whereby the operation of one control or mechanism brings about or prevents the operation of another.

Joint motion: a method for coordinating the movement of the joints, so that all joints arrive at the desired location simultaneously.

Limiting device: a device that restricts the maximum envelope (space) by stopping or causing to stop all robot motion, independent of the control program and the application programs.

Maintenance: the act of keeping the robots and robot systems in their proper operating condition.

Mobile robot: a self-propelled and self-contained robot capable of moving over a mechanically unconstrained course.

---

[1] Terms and definitions listed in the following are from *OSHA's TED 1.15, September 22, 1995, Office of Science and Technology*.

Muting: the deactivation of a presence-sensing safeguarding device during a portion of the robot cycle.

Operator: the person designated to start, monitor, and stop the intended productive operation of a robot or robot system. An operator may also interface with a robot for productive purposes.

Pendant: any portable control device, including teach pendants, that permits an operator to control the robot from within the restricted envelope (space) of the robot.

Presence-sensing safeguarding device: a device designed, constructed, and installed to create a sensing field or area to detect an intrusion into the field or area by personnel, robots, or other objects.

Program:

- (noun) a sequence of instructions to be executed by the computer or robot controller to control a robot or robot system;
- (verb) to furnish (a computer) with a code of instruction;
- (verb) to teach a robot system a specific set of movements and instructions to accomplish a task.

Rebuild: to restore the robot to the original specifications of the manufacturer, to the extent possible.

Remanufacture: to upgrade or modify robots to the revised specifications of the manufacturer and applicable industry standards.

Repair: to restore robots and robot systems to operating conditions after damage, malfunction, or wear.

Robot manufacturer: a company or business involved in the design, fabrication, or sale of robots, robot tooling, robotic peripheral equipment or controls, and/or associated process ancillary equipment.

Robot system integrator: a company or business that either directly or through a subcontractor will assume responsibility for the design, fabrication, and integration of the required robot, robotic peripheral equipment, and other required ancillary equipment for a particular robotic application.

Safeguard: a barrier guard, device, or safety procedure designed for the protection of personnel.

Safety procedure: an instruction designed for the protection of personnel.

Sensor: a device that responds to physical stimuli (such as heat, light, sound, pressure, magnetism, motion, etc.) and transmits the resulting signal or data by providing a measurement, operating a control, or both.

Service: to adjust, repair, maintain, and make fit for use.

Single point of control: the ability to operate the robot so that initiation or robot motion from one source of control is possible only from that source, and cannot be overridden from another source.

Slow speed control: a mode of robot motion control where the velocity of the robot is limited to allow persons sufficient time either to withdraw the hazardous motion or stop the robot.

Start-up: routine application of drive power to the robot or robot system.

Start-up, (initial): initial drive power application to the robot or robot system after one of the following events:

- Manufacture or modification
- Installation or reinstallation
- Programming or program editing
- Maintenance or repair

Teach: the generation of storage of a series of positional data points effected by moving the robot arm through a path of intended motions.

Teach mode: the control state that allows the generation and storage of positional data points effected by moving the robot arm through a path of intended motions.

Teacher: a person who provides the robot with a specific set of instructions to perform a task.

Tool center point (TCP): the origin of the tool coordinate system.

User: a company, business, or person who uses robots and who contracts, hires, or is responsible for the personnel associated with robot operation.

Summary

Mark Twain said, "There are lies, damned lies, and statistics," thereby giving statistics a bad name. While we know statistics can be manipulated to deceive, we also recognize them for the invaluable tool they are. We use on-the-job injury statistics as a benchmark, a way to gauge the success of workplace safety, both specifically—in our own workplace, and generally—in workplaces across the nation.

We want to see those numbers dropping—and we especially want to know why some numbers are still high. We applaud the reduction of injuries and fatalies due to safe work practices and training, but we know also that on the average, 153 people die every day from workplace-related illness (137) or injury (16). Proper machine guarding is one way to reduce that statistic.

Cited References

Brubaker, S. A., Optimum Machine Guarding. *Occupational Health and Safety.* December, 1997.

Ferry, T., *Safety and Health Management.* New York: Van Nostrand Reinhold, 1990.

Karr, Al., How Far Have We Come? *Safety and Health*, January, 1999.

LePree, J., *Complying with OSHA's Guarding Standards can avert Accidents.* Cahners Business Information, 1998.

CHAPTER 2

MECHANICAL AND NON-MECHANICAL HAZARDS OF UNGUARDED MACHINES

It is important to be safety conscious when dealing with any job that requires the use of machinery. Statistics show that the majority of machinery-related accidents occur as the result of human negligence. Errors include: taking shortcuts to save time; failure to read the operator's manual; ignoring a warning; improper or lack of instruction; failure to follow safety rules; and improper or lack of maintenance.

— Margentino, M. R., and Malinowski, K., 1992.

Ed climbed back in the seat and tried the lever. He pushed down hard, and only got hissed at by the pump. He pulled up, and got the same treatment. "Damn-dead-stuck-bucket-a-rust."

Ed remembered climbing back down and looking at the hydraulic pump over the top of the truck frame. He didn't recall making the decision not to, but somehow he didn't want to lie back down in the mud again. He couldn't remember if he was thinking about it at all. He couldn't have been or he wouldn't have done it. But what he could remember, and what his dreams would never let him forget, was watching his hand reach out with the pipe wrench between the frame and the stuck dump bed, and how it was the last time he would ever see his right arm.

— Bodett, Tom, *The Big Garage on Clear Shot*, p 179, 1990.

2.1 INTRODUCTION

Methods of machine guarding vary greatly, depending on the machine and the types of hazards it presents.

The above statement seems simple enough to comprehend—and it is. Obviously, industry uses many different types of machines, run by workers. A wide variety of available machines means a wide variety of machine hazards or potential hazards. These hazards are the focus of this chapter, our intent, to familiarize you with the hazards of unguarded machines.

2.2 HAZARDS OF UNGUARDED MACHINES

Two types of hazards are associated with unguarded machines. The first type is mechanical—hazards arising from the motion or operation of the machine. We present examples of these machine motions and actions in Figure 2.1. The other type of hazard is one unrelated to the movement of the machine.

2.2.1 Mechanical Hazards

Machine *mechanical hazards* are a result of machine motion or machine action (see Figure 2.1 and 2.2). Figure 2.2 shows five types of machine motion/action that create machine hazards:

rotation, reciprocating or transverse motion, in-running nip points or pinch points, cutting actions, and punching, shearing, and bending.

Note: Specific processes performed by the machine can also present hazardous conditions. Examples include cutting, punching, shearing, and bending. The location of these actions is called the point-of-operation (see Sections 2.2.1.4 and 2.2.1.5).

2.2.1.1 Rotation

A variety of mechanical motions and actions may present hazards to the worker. Two hazards are presented by *rotating motions*. Protruding objects (such as bolts, screws, and burrs, rough surfaces and the rotating motion itself) all tend to take up, grab, or seize hair and clothing, pulling it (and whatever it is connected to) into the machine. Examples of rotating motion are the spindle, chuck and bit of a drill motor, shafts, splines, couplings, feed stock, flywheels and handwheels. Additionally, set screws, burrs, rough surfaces, and rotary or eccentric motion all tend to grab clothing and hair. Smooth, slowly rotating surfaces can also catch clothing, hair or other loose material and can cause injury. The hazard is obvious: once caught, clothing or hair can quickly pull a person into a machine. Keep in mind that skin contact with these moving surfaces can also produce injury.

The second type of hazard is called an *in-running nip point*. In-running nip point hazards are caused by the rotating parts of machinery, created when machine parts rotate towards each other,

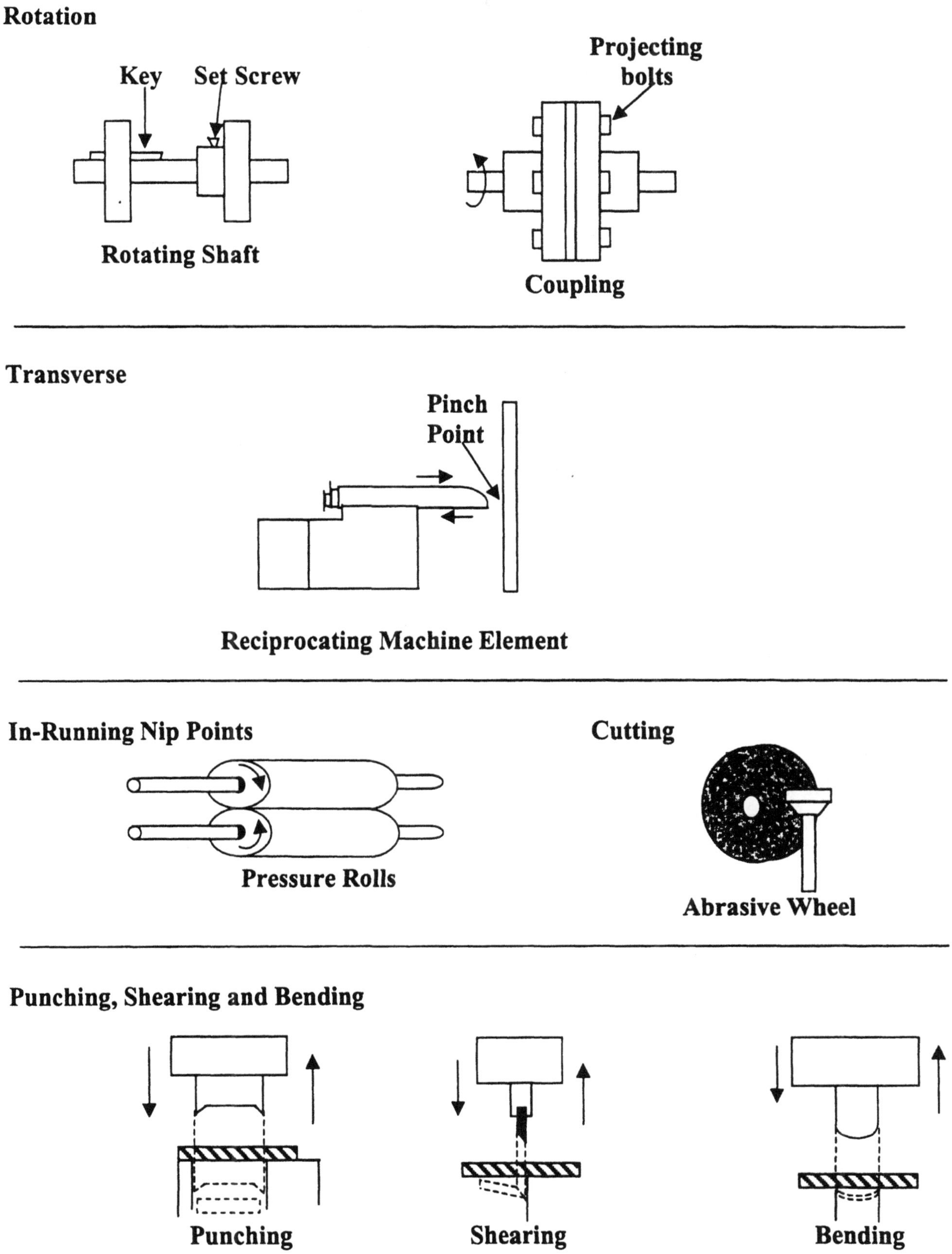

Figure 2.1. Examples of five types of machine motions.

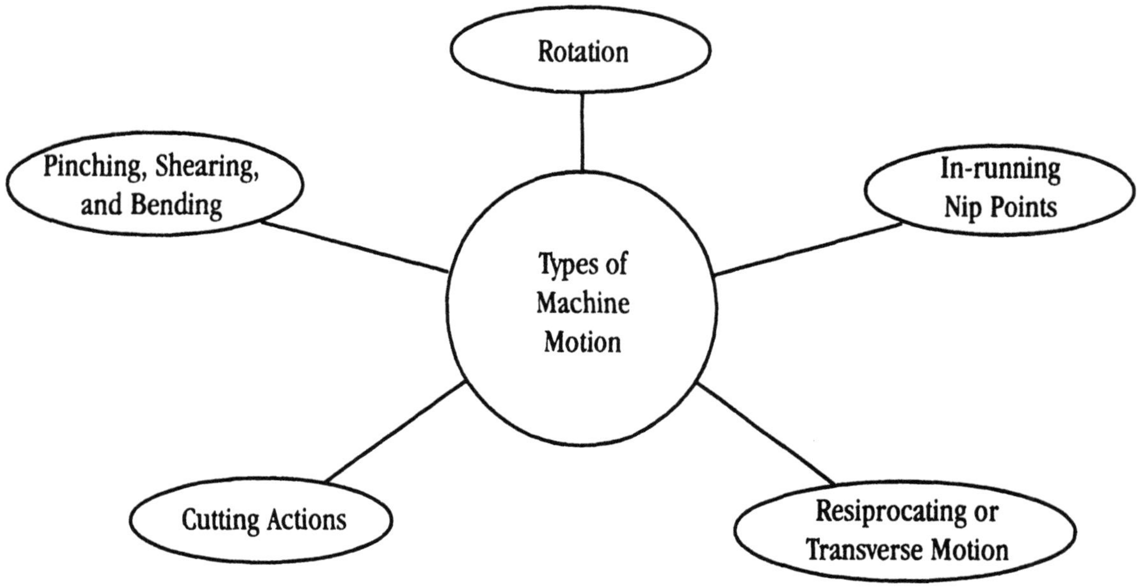

Figure 2.2. Types of machine motions.

or when a rotating part rotates toward a stationary object (which can create a shearing, crushing, or abrading action—e.g., spoke handwheels, or flywheels). Nip points can present a severe hazard by pulling objects into the rotating parts, or crushing the object against the stationary component. This danger is common on machinery with intermeshing gears, rolling mills, and calendar. Specific machine parts that provide in-running nip points include pressure rolls, chains and sprockets, spoked handwheels, and rack and pinion (see Figure 2.1).

2.2.1.2 Reciprocating Motions

Reciprocating motions may be a hazard because, during the back-and-forth or up-and-down motion, a worker may be struck by or caught between a moving and a stationary part (see Figure 2.1). Someone could also get pinched or sheared between fixed and moving parts of the machine, or between the part of the machine and a surrounding fixed object. The reciprocating motion of an automated machine equipped with a moving part (such as a moving table) is an example. In operation, the table moves back and forth while processing the material. The hazard, obviously, is from being struck by the moving table.

2.2.1.3 Transverse Motions

Transverse motion (movement in a straight, continuous line, such as a conveyor belt) creates a hazard similar to those of reciprocating motions, because a worker may be struck or caught in a pinch or shear point by the moving part (see Figure 2.1).

2.2.1.4 Cutting Action

Cutting action may involve rotating, reciprocating, or transverse motion. Simply stated, a cutting action can be defined as one that cuts or removes material from an object. The danger of the cutting action exists at the point of operation where finger, arm, and body injuries can occur, and where flying chips or scrap material can strike the head, particularly in the area of the eyes or face. Common machines that exhibit cutting action include abrasive grinding wheels, band saws, lathes, milling machines, drills, circular saws, and planers. Obviously, if the machine can cut leather, metal, paper, or wood, it can easily cut human tissue. Such hazards are present at the *point of operation* (see Figure 2.1).

2.2.1.5 Punching, Shearing, and Bending

Punching action results when power is applied to a slide (ram) for the purpose of blanking, drawing, or stamping metal or other materials. This type of action describes two machine components that come together (both may move or one may be stationary). Typical machines used for punching operations are power presses and iron workers.

Shearing action involves applying power to a slide or knife to trim or shear metal or other materials. A hazard occurs at the point of operation—where stock is actually inserted, held, and withdrawn. Examples of machines used for shearing operations are mechanically, hydraulically, or pneumatically powered shears (flying shears).

Bending action results when power is applied to a slide to draw or stamp metal or other materials. A hazard occurs at the point of operation where stock is inserted, held, and withdrawn. Equipment that uses bending action includes power presses, press brakes, and tubing benders (see Figure 2.1).

2.2.2 Nonmechanical Hazards

While this text concentrates on concepts and techniques for safeguarding mechanical motion, nonmechanical hazards can also result from unguarded machinery. These include electrical power sources, high pressure systems, chemical emissions, noise, and contact with flying objects (see Figure 2.3).

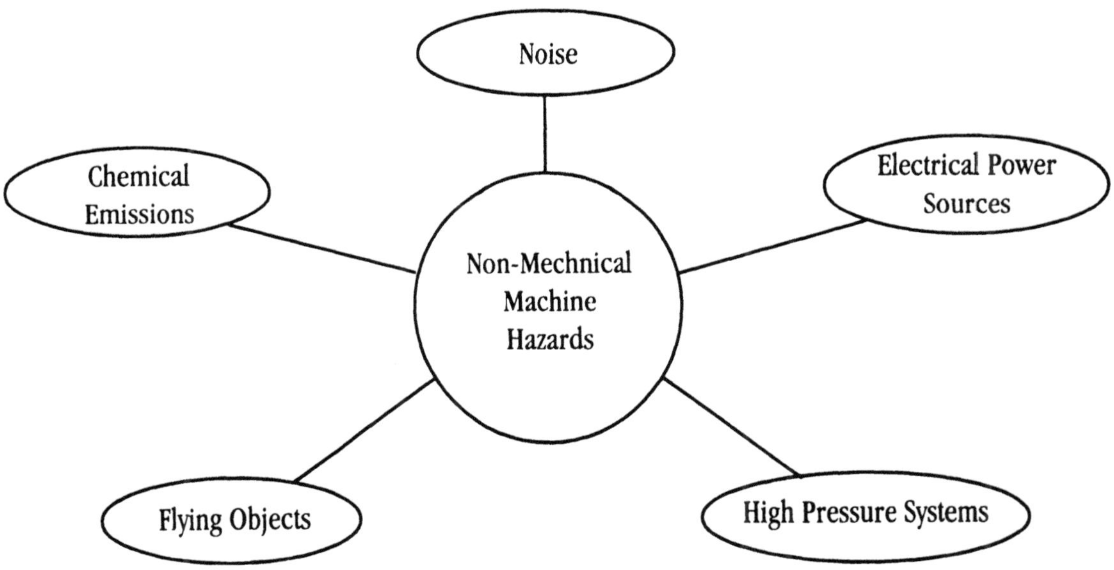

Figure 2.3. Non-mechanical machine hazards.

2.2.2.1 Electrical Power Sources

All machine power sources are potential sources of danger. Contact with improperly grounded sources or damaged wiring can cause severe injury or death.

> *Note: When using electrically powered or controlled machines, the equipment as well as the electrical system itself must be properly grounded.*

Replacing frayed, exposed, or old wiring will also help to protect the operator and others from electrical shocks or electrocution. Standards for electrical power sources can be found in the National Electrical Code (NEC) and from the American National Standards Institute (ANSI).

2.2.2.2 High Pressure Systems

High pressure systems require careful inspection and maintenance to prevent possible failure from pulsation, vibration, or leaks. Such failures can cause, among other things, the release of pressurized gases, gauge blowout, and line rupture, producing explosions and flying objects.

2.2.2.3 Chemical Emissions

Various chemicals such as cutting fluids, coolants, coating, cleaning fluids, and other potentially harmful substances are used with certain machines. Because many of these chemical substances are potentially hazardous (many of these substances can cause ailments ranging from dermatitis to serious illnesses and disease), operators may need protection. For instance, large cutting machines typically use nebulizers that atomize a fine spray of lubricant at the worker's face. In many cases, barrier guarding or shielding (along with ventilation and personal protective equipment and clothing) can assist in protecting the worker from exposure to mists.

> *Note: When providing workers with personal protective equipment (PPE) for protection from chemical emissions resulting from machine use/operation, note that PPE is a temporary solution—such hazards should be controlled or eliminated (engineered out of) from the workplace.*

2.2.2.4 Noise

Noise (defined as any unwanted sound) is an inherent hazard—a by-product of machine operation. Machine noise can result in a number of hazards to workers. Noise can startle and disrupt concentration and can interfere with communications, thus hindering the worker's safe job performance. Research has linked noise to a whole range of harmful health effects, from hearing loss and aural pain to nausea, fatigue, reduced muscle control, and emotional disturbance. Properly designed and installed engineering controls (such as the use of sound-dampening materials and other guards) can often assist in reducing noise levels.

> *Note: Remember that improperly installed guards can actually increase noise levels from vibration, or can channel the noise directionally like a speaker.*

Personal protective equipment (PPE), such as ear plugs and muffs, can also help control the harmful effects of noise. Additionally, administrative controls that involve removing the worker from the noise source can be an effective measure when feasible.

2.2.2.5 Flying Objects

Machine guarding is often used to act as a barrier between the worker and the machine, especially when the machine produces flying objects such as wood from cutting, metal trimmings from shearing, and objects which become entrained in rotating machines—all of which present severe hazards to workers.

Summary

"You can lead a horse to water, but you can't make him drink" is an old cliché. Likewise, you can provide workers with machine guarding, but use depends in part on ease of use and perceived risk. Any guard a worker finds troublesome to work around can be circumvented, especially if the worker doesn't believe, deep down, that he or she needs the protection.

Guard design and operation should work to protect the worker—but only good design, training, and management support will ensure that workers use the guards consistently and properly.

Cited Reference

Bodett, Tom, *The Big Garage on Clear Shot.* William Morrow and Company, New York, 1990.

Margentino, M. R., and Malinowski, K., *Operation of Farm Machinery.* Document FS620. New Jersey: Rutgers, State University of New Jersey, 1992.

CHAPTER 3

GUARDS

A recent OSHA citation for a safety violation:[2]

> ***"No Guard on Saw, Worker Cuts Off Thumb, Fingers"***
>
> *Company: Kitchen Design Company*
>
> *Business: Manufacturer of wooden kitchen cabinets*
>
> *Why Inspected? Investigation into incident when worker cut off part of his thumb and two fingers while using an unguarded table saw*
>
> *Fine: $46,514*
>
> *Reasons for fine:*
>
> *Two willful citations ($37,500):*
>
> - Allowing workers to operate table saw with an unguarded 10-inch blade
> - Table saw lacked devices to prevent the blade from being kicked back at the worker.
>
> *Thirteen serious citations ($9,014):*
>
> - No written emergency action plan
> - Improperly stored flammable material
> - Failure to enforce use of protective eyewear
> - Unguarded machine blades
> - Ungrounded electrical equipment
> - Failure to establish a written energy control program.

---

[2] From *Safety Compliance Alert*, p. 4, December 23, 1997.

3.1 INTRODUCTION

When machine safeguarding is needed, we can consider several safeguarding methods, including guards, devices (see Chapter 4), feeding and ejecting methods (see Chapter 5), and location and distance (see Chapter 6). This chapter provides an overview of guards, types of guards, and their advantages and disadvantages. However, before we begin a discussion about guards, we first provide a general discussion of guard requirements—for any machine guard type.

3.1.1 Requirements of Machine Guards

OSHA's 3067 *Concepts and Techniques of Machine Safeguarding* (1992—Revised) asks the question, "What must a safeguard do to protect workers against mechanical hazards?" The answer? Safeguards must meet these minimum general requirements (see Figure 3.1):

Prevent contact: The safeguard must prevent hands, arms, and any other part of a worker's body from making contact with dangerous moving parts. A good safeguarding system eliminates the possibility of the operator or other workers placing parts of their bodies near hazardous moving parts.

Secure: Workers should not be able to easily remove or tamper with the safeguard, because a safeguard that can easily be made ineffective is no safeguard at all. Guards and safety devices should be made of durable material that will withstand the conditions of normal use. They must be firmly secured to the machine.

Protect from falling objects: The safeguard should ensure that no objects can fall into moving parts. A small tool dropped into a cycling machine could easily become a projectile that could strike and injure someone.

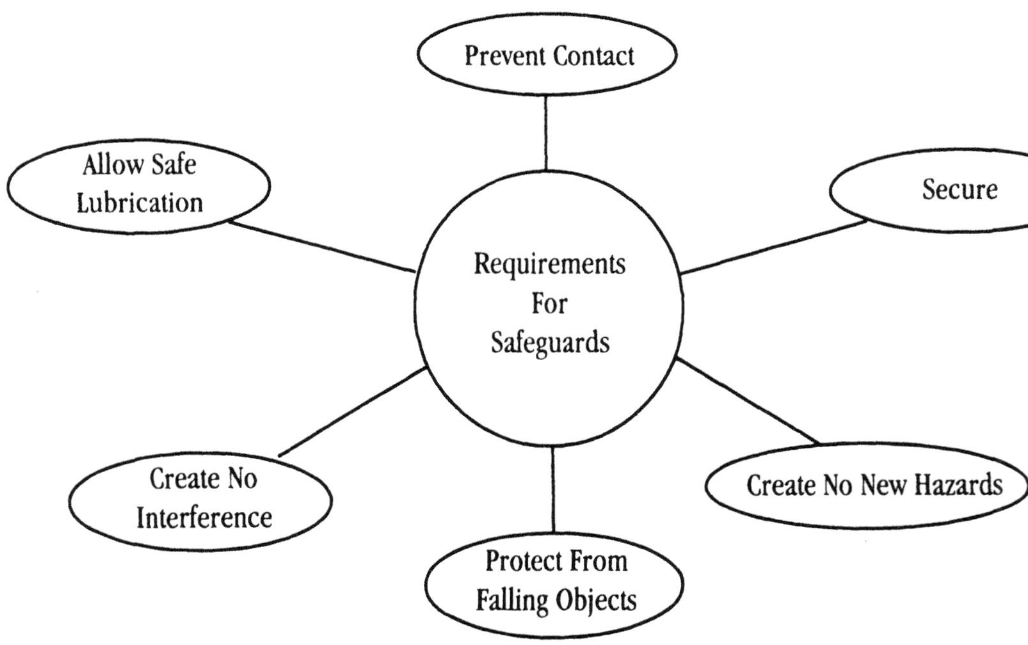

Figure 3.1. Safeguard requirements.

Create no new hazards: A safeguard defeats its own purpose if it creates a hazard of its own such as a shear point, a jagged edge, or an unfinished surface that can cause a laceration. The guard edges, for instance, should be rolled or bolted in such a way that they eliminate sharp edges.

Create no interference: Any safeguard that impedes a worker from performing the job quickly and comfortably might soon be overridden or disregarded. Proper safeguarding can actually enhance efficiency, since it can relieve the worker's apprehensions about injury.

Allow safe lubrication: If possible, one should be able to lubricate the machine without removing the safeguards. Locating oil reservoirs outside the guard with a line leading to the lubrication point will reduce the need for the operator or maintenance worker to enter the hazardous area.

3.1.2 Guard Openings

In the last statement above—allow for safe lubrication—we make the point that guards cannot (in all cases) be expected to totally encapsulate a machine without allowing for routine maintenance and for operation. Guard openings must also be provided to allow the insertion of materials into a machine for processing, and to monitor machine operation and action. Obviously, the larger the opening, the farther one can reach through it. This opening-reach relationship is important and has been the subject of several research studies by the manufacturing industry, the insurance industry and safety engineering professionals. As a result of these studies, the data in Table 3.1 was developed to define how far a barrier must be placed from a danger point for certain size guard openings.

Table 3.1. Distances a Guard Must be from a Danger Line

Distance of Opening from Danger Line (in.)	Maximum Width of Opening (in.)
.5 to 1.5	0.25
1.5 to 2.5	0.375
2.5 to 3.5	0.5
3.5 to 5.5	0.625
5.5 to 6.5	0.75
6.5 to 7.5	0.875
7.5 to 12.5	1.25
12.5 to 15.5	1.5
15.5 to 17.5	1.875
17.5 to 31.5	2.125

A *danger line* or *point* is where the dangerous action of a machine occurs, or a location that narrows to 0.375 in., which forms a pinch point. Figure 3.2(a) illustrates the danger line for openings between rollers, and the application of distance-opening data. For rollers with openings greater than 0.375 in., the fingers and hands may not sustain severe crushing. Not only are straight line distances important, but so is the potential for reaching around or over some obstacles (Brauer, p. 148, 1994). Figure 3.2(b) gives an example.

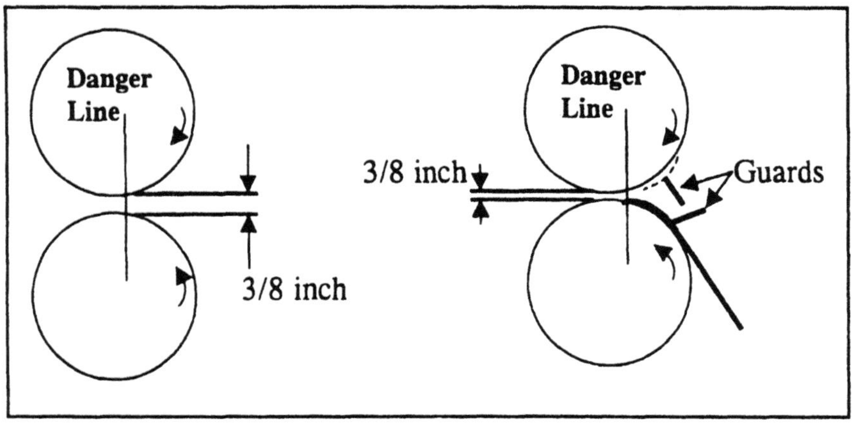

Figure 3.2(a). The danger line for in-running rolls. Guard placement is determined from this defined location.

Figure 3.2(b). Example of using distance to keep someone from reaching into dangerous parts of a machine. The application and various standards determine whether the use of distance provides sufficient protection.

3.1.3 Guard Construction

Various standards give recommended materials for construction. However, one thing is certain: Guards must be substantial. Today, many builders of single-purpose machines provide point-of-operation and power transmission safeguards as standard equipment. However, not all machines in use have built-in safeguards provided by the manufacturer.

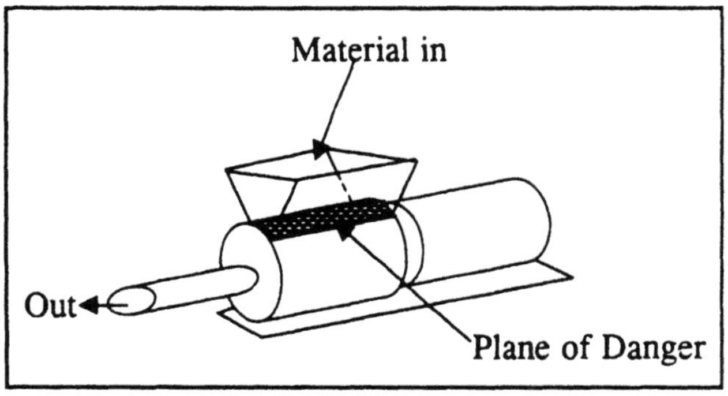

OSHA 3067 (1992—revised) points out that guards designed and installed by the builder offer two main advantages:

- They usually conform to the design and function of the machine.
- They can be designed to strengthen the machine in some way, or to serve some additional functional purposes.

User-built guards are sometimes necessary for a variety of reasons. They have the following advantages:

- Often with older machinery, they are the only practical safeguarding solution.
- They may be the only choice for mechanical power transmission apparatus in older plants, where machinery is not powered by individual motor drives.
- They permit options for point-of-operation safeguards when skilled personnel design and make them.
- They can be fashioned to fit unique and ever-changing situations.
- They can be installed on individual dies and feeding mechanisms.
- Design and installation of machine safeguards by plant personnel can help to promote safety consciousness in the workplace.

They also have disadvantages, however:

- User-built guards may not conform well to the configuration and function of the machine.
- User-built guards are sometimes poorly designed or built.

Note: Point-of-operation safeguarding is complicated by the number and complexity of machines and also by different uses for individual machines. For these reasons, not all machine builders provide point-of-operation guards on their products. Therefore, a point-of-operation guard (in many cases) can only be made and installed by the user after a thorough hazard analysis of the work requirements. The important point to remember is that a poorly designed, built, or installed guard may create a hazard rather than eliminate one. To be effective, the guard must safeguard the employee—while allowing the work to continue with minimum disruption to the production process.

A significant difference between power transmission guards and point-of-operation guards is that the former type needs no opening for feed stock (see Figure 3.3). The only openings necessary for power transmission guards are those for lubrication, adjustment, repair, and inspection. These openings should be provided with interlocked covers that cannot be removed except by using tools for service or adjustment.

Note: To be effective, power transmission guards should cover all moving parts so that no part of the operator's body can come in contact with them (see Figure 3.3).

Earlier, we stated that guards should be substantial. In most circumstances, metal is the best material for guards (and usually is substantial). Guard framework is generally made from structural shapes, pipe, bar, or rod stock. Filler material generally is expanded or perforated solid sheet metal, or wire mesh. Using plastic or safety glass may be feasible where visibility is required.

Figure 3.3. Examples of user-built power transmission guards.

Note: Guards made of wood are not recommended because of their flammability and lack of durability and strength. However, in areas where corrosive chemicals are present, wooden guards may be the better choice.

3.2 TYPES OF GUARDS

Figure 3.4 shows the four general types of guards: fixed, interlocked, adjustable, self-adjusting, and special guards. In this section, we discuss each type of guard.

3.2.1 *Fixed Guards*

As its name implies, a *fixed guard* is a permanent machine attachment (see Figure 3.3). Fixed guards are not dependent upon moving parts to perform their intended function. They also do not move to accommodate the work being performed. This type of guard often has an access panel or transparent portions to view the operation of the machine. It may be constructed of sheet metal, screen, wire cloth, bars, plastic, or any other material that is substantial enough to withstand whatever impact it may receive, and to endure prolonged use. If its application is feasible, this guard is usually preferable to all other types because of its relative simplicity, as well as its limited means of access and removal (permanence). Fixed guards also provide the maximum protection, prevent access to dangerous parts of the machine, need minimal maintenance, and are suitable to many applications. Examples include a fixed guard on a power press whereby the stock is fed through the side of the guard into the die area, with the scrap stock exiting on the opposite side. Fixed guards are also used as shields on belt and pulley (power transmission) systems (see Figure 3.3), band saws, and in many other applications. Disadvantages of fixed guards include limited visibility and the need for additional safeguards when opened or removed for maintenance of the machine.

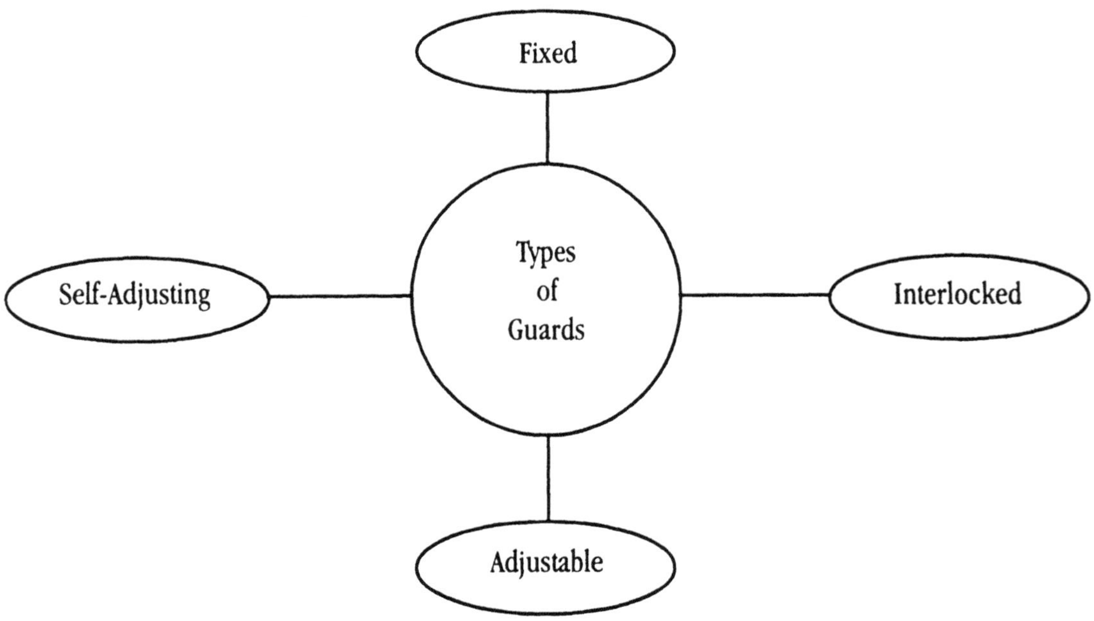

Figure 3.4. Types of Guards.

3.2.2 Interlocked Guards

The purpose of a safety interlock is to ensure that a machine will only operate when all safety guards and gates are correctly positioned. During operation, when an interlock circuit opens, the machine will revert to fail-safe operation and stop.

Probably the most familiar type of *interlock guard* is the interlock system installed on home clothes dryers and industrial drying tumblers. When the door of the machine is opened, the circuit is broken and the revolving cylinder stops; the machine is immediately made inoperable. In industry, these guards are used in situations that require operators to have frequent access to the machine for setup, adjustment, or maintenance purposes.

An interlocked guard may use electrical, mechanical, hydraulic, pneumatic power, or any combination of these. Interlocks should not prevent "inching" by remote control if required. Replacing the guard should not automatically restart the machine. To be effective, all movable guards should be interlocked to prevent occupational hazards.

3.2.2.1 A Closer Look at Interlocks[3]

The National Fire Protection Association (NFPA), in its NFPA 79—*Electrical Standard for Industrial Machinery*, classifies machine stopping in three different categories: 0, 1, and 2.

> *Category 0* defines an uncontrolled stop of a machine. In this case, stopping a machine's motion is accomplished by removing power to the machine actuators; all brakes or other mechanical stopping devices are actuated.
>
> > *Note: an actuator is that part of a machine actuating system to which an external actuating force is applied (e.g., a handle, knob, push-button, roller, or plunger).*
>
> *Category 1* specifies a controlled stop with power applied to the machine actuators to cause them to stop (usually applied by very large motors and machines). Once stopped, the power is then removed from the actuators.
>
> > *Note: NFPA defines a* controlled stop *as the stopping of the machine motion by reducing the command signal to zero, but power to the machine actuators is still available.*
>
> *Category 2* describes a controlled stop as above, but in this case, power is still available to the machine actuators after the stop.
>
> > *Note: NFPA requires that "each machine should be equipped with a Category 0 stop. Category 1 and Category 2 stops shall be provided where required by the safety and functional requirements of the machine. Category 0 and Category 1*

[3] Adapted from Puddicombe's "A Guide to Using Safety Interlock Switches. *Occupational Hazards*. pp. 45-50, March, 1998.

stops shall be functional regardless of operating mode, and a Category 0 stop shall take priority. Stop functions shall operate by deenergizing the circuit and shall override related start functions."

Various U.S. safety codes rate guarding devices based on mechanical design and wiring, including the wiring or "logic" of their safety circuits. Safety interlocks fall into two separate categories: positive mode/direct contact, and positive mode/noncontact.

Positive mode/direct contact (sometimes referred to as direct operation) works with a normally closed (NC) limit switch. A mechanical device or interlock breaks the circuit when the safety gate or guard is disengaged.

Positive mode/noncontact type safety interlock is similar to a direct contact device, but instead relies on a sensing field (magnetic, capacitive, or proximity) switching device integrated into the safety interlock.

Even with safety interlock systems in place, a failure of one of its parts may result in an unsafe condition. This unsatisfactory situation is remedied by the use of redundant device outputs such as those used in conjunction with a Safety Rated Relay Module (SRRM). The SRRM provides safety-related redundant output for use as an Emergency Stop Relay (ESR) or as a Master Control Relay (MCR) for terminating the end control element's motion. An SRRM incorporates three elements of safety in its design: input redundancy, output redundancy and restart protection.

Input redundancy refers to a device with two input channels. They can handle one of two hard-wired device strings. They may be redundant devices or individual safety-related devices.

Output redundancy provides wiring anywhere from 2 to 10 output channels to an end element control device that can include motor starters or solenoid valves.

Restart prevention is a safety application whereby if either output relay fails to deenergize, the SRRM will not allow the machine to restart or perform a successive cycle.

Interlocks have the advantages of providing a high degree of protection, and providing quick and easy access to the machine. The disadvantages associated with interlocks include the requirement to perform frequent maintenance and adjustment, and that they can be made inoperable.

3.2.3 Adjustable Guards

Adjustable guards are useful because they allow flexibility in accommodating various sizes of stock; they can be adjusted by the operator when the point-of-operation must be accessible because of varying sizes of work. Horizontal band saws, power presses, routers, shapers, table saws, and several other machines typically are equipped with adjustable guards. An example of an adjustable guard is shown Figure 3.5.

The advantages of adjustable guards are they make varying sizes of work accessible to the point-of-operation, and they have a wide variety of applications. The disadvantages include that the degree of protection is dependent on correct adjustment by the operator; they offer no protection during adjustment; they require frequent adjustment and maintenance; and they can limit visibility.

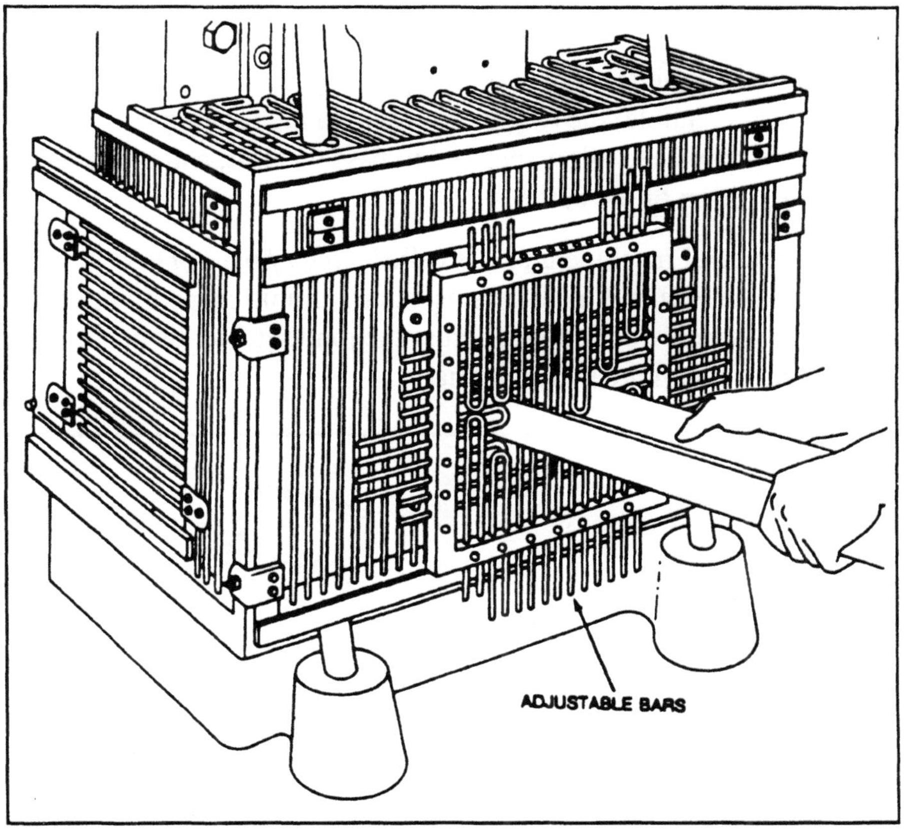

Figure 3.5. Example of an adjustable guard (from OSHA 3067).

3.2.4 Self-Adjusting Guards

When a *self-adjusting guard* is employed, the opening of this type of barrier is determined by the movement of the stock. As the operator moves the stock into the danger area, the guard is pushed away, providing an opening that is only large enough to admit the stock. After the stock is removed, the guard returns to the rest position. This guard protects the operator by placing a barrier between the danger area and the operator. The guards may be constructed of plastic, metal, or other substantial material. Self-adjusting guards offer different degrees of protection.

> *Note: The key advantage of self-adjustable guards is that they do not require the operator to adjust them. That is, as the piece of work is inserted, the guard is pushed away, leaving an opening only large enough to accommodate the work piece.*

Self-adjusting guards are typically found on radial arm saws, table saws, circular saws, and jointers. Disadvantages of adjustable guards include a lower degree of protection; limited visibility; and required frequent adjustment and maintenance.

3.2.5 Special Guards

Brauer (1994) points out that many special guards are unique to particular machines or machine actions, for example, the *ring guard* used on the rotating cutter on a shaper (see Figure 3.6). The ring guard adjusts to allow material to pass under it, and provides protection from the remaining exposed portion of the cutter.

Another special machine guard is called a *leg of mutton guard,* used to cover the cutting head of a jointer (see Figure 3.6). It gets its name from its shape. In operation, it swings horizontally out of the way when wood moves across the jointer table. Its position varies with adjustments made in the fence that guides the material.

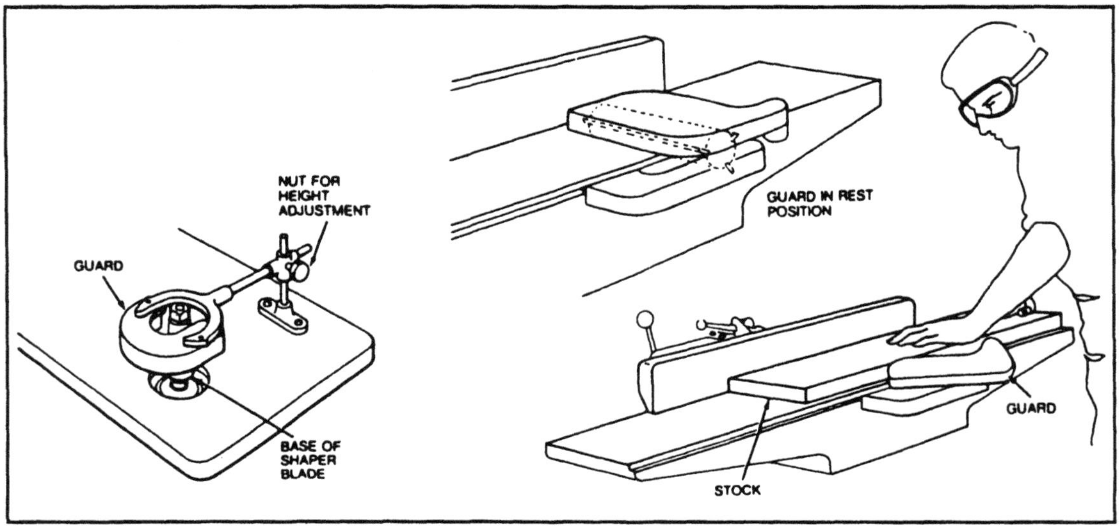

Figure 3.6. Examples of some special guards; (a) ring guard; (b) leg of mutton guard (from OSHA 3067).

> *Note: Because the leg of mutton guard can only move horizontally, it will not protect all jointer cuts. Guards that float vertically give protection when material moves over the cutting head and extends over the side of the jointer table.*

A special guard called a *hood guard* is used on circular saws. The guard moves or "floats" vertically (on hand-held machines) as material is moved into the blade.

Probably the best known and easiest to recognize special fixed guards (for bench-top mounted machines; floating special guards for portable hand-held grinders) are those used on *grinding wheels* or *abrasive wheels* (see Figure 3.7). Keep in mind that because many kinds of grinding operations and grinding wheels are used, many variations and additional design features have been designed to fit each variation. However, the basic operation of grinding wheel systems and the function of their guards are similar. The guards help keep operators from coming into contact with the wheel itself, and they help contain fragments should a wheel fracture, disintegrate, and fly apart.

Note: Remember that a "ring" test must be performed on abrasive wheels before they are put into service. Also note that grinding wheel machines (such as the one shown in Figure 3.7) are typically included on OSHA's list of most often cited non-compliance violations. Usually these citations are based on lack of guards, or on guards being broken or inadequate for their intended purpose.

Figure 3.7. Examples of special guards used on table-mounted grinding wheels.

Table 3.2 summarizes the safeguarding action, advantages, and limitations of the four major types of guards—fixed, interlocked, adjustable, and self-adjusting.

Summary

OSHA's minimum requirements for machine guarding (prevent contact, secure, protect from falling objects, create no new hazards, create no interference, allow safe lubrication) provide a good foundation—an effective starting point for machine guarding. But they are only the foundation. Protecting workers from the hazards presented by moving machinery involves more than these six points. Identifying the hazards to guard against in your workplace is the next essential step.

Cited References

Brauer, R. L. *Safety and Health for Engineers*. New York: Van Nostrand Reinhold, 1994.

Table 3.2. Safeguarding Action – Advantages and Limitations

METHOD	SAFEGUARDING ACTION	ADVANTAGES	LIMITATIONS
Fixed	Provides a barrier	• Suitable for many applications • In-house fabrication • Provides maximum protection • Usually requires minimum maintenance • Suitable to high production, repetitive operations	• Interferes w/visibility • Limited to specific operations • Machine adjustments and repair often require its removal, thereby necessitating other means of protection for maintenance personnel
Interlocked	Shuts off or disengages power and prevents starting of machine when guard is open; should require the machine to be stopped before the worker can reach into the danger area	• Can provide maximum protection • Allows access to machine for removing jams without time consuming removal of fixed guards • May be easy to disengage jams	• Requires careful adjustment & maintenance
Adjustable	Provides a barrier that may be adjusted to facilitate a variety of production operations	Can construct to suit many specific applications • Adjustable to admit varying sizes of stock	• Hands may enter danger area—protection may not be complete at all times • May require frequent maintenance and/or adjustment • The guard may be made ineffective by the operator • May interfere with visibility
Self-adjusting	Provides barrier that moves according to size of stock entering the danger area	Off-the-shelf guards often commercially available	• Does not always provide maximum protection • May interfere with visibility • May require frequent maintenance & adjustment

Source: OSHA 3067.

CHAPTER 4

DEVICES

IDENTIFYING SAFETY HAZARDS

A preliminary analysis of safety risks, even if subjective, will allow the designer to categorize the required safety control system. Without a specific reference in an existing code, this becomes the basic foundation for safety analysis. The first step involves performing a machine hazard analysis to:

- *Identify points of operation*
- *Identify each specific hazard*
- *Develop a description of each hazard.*

The next step in risk analysis considers four areas of safety concern:

- *First, determine the possibility of avoiding injury*
- *Next, determine the severity of the injury*
- *Estimate the number of times someone will be exposed to the hazard over a given period of time*
- *Estimate the likelihood that an accident will occur.*

The third and final step in the hazard analysis is to determine which category of safety control system to apply. This is based on reliability requirements, system interlock, and the safety rating of the hardware.

— Puddicombe, R., p. 50, 1998.

4.1 INTRODUCTION

Along with guards, certain *devices* are also used to safeguard machinery. A safety device may perform one of several functions. It may:

- stop the machine if a hand or any part of the body is inadvertently placed in the danger area
- restrain or withdraw the operator's hands from the danger area during operation
- require the operator to use both hands on machine controls, thus keeping both hands and body out of danger
- or provide a barrier which is synchronized with the operating cycle of the machine to prevent entry to the danger area during the hazardous part of the cycle.

Point-of-operation devices are typically employed whenever providing an adequate guard is not possible. Point-of-operation guards ensure that the machine cannot operate or create a hazard while an operator has fingers or hands in the point-of-operation.

Note: Simply stated, a machine-guarding device inhibits normal operation of a machine if any portion of a person's body is within a hazardous area.

4.2 TYPES OF DEVICES

In this section, we discuss the types of safeguarding devices (controls or attachments), including presence sensing devices, pullback mechanisms, restraints, safety trip controls, and gates (see Figure 4.1).

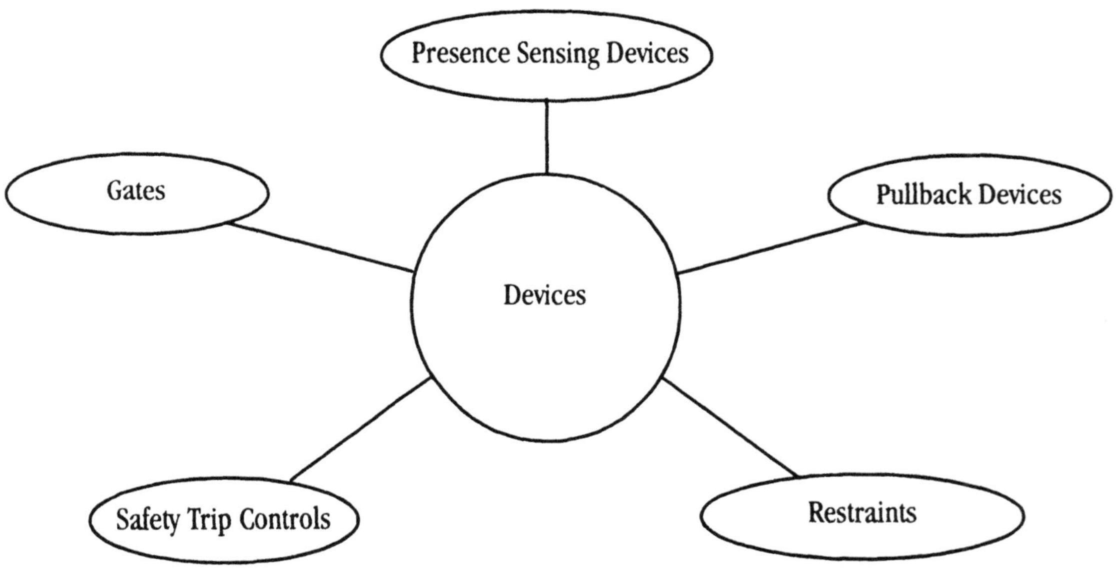

Figure 4.1. Devices used for safeguarding machines.

4.2.1 Presence-Sensing Devices

Presence-sensing devices typically operate on photoelectric, radiofrequency, or electromagnetic principles, to disengage the machine when something is detected in the hazardous area. The photoelectric (optical) presence-sensing device uses a system of light sources (light beams) and sensing mechanisms (controls or sensors), which can interrupt the machine's operating cycle. An array of beams forms a horizontal, vertical or combined pattern. If the light field is broken, the machine stops and will not cycle. Because the array is not easily seen, the operator must depend on a display to know if it is functioning correctly. This device must be used only on machines that can be stopped before the worker can reach the danger area. The design and placement of the guard depends upon the time needed to stop the mechanism, and the speed at which the employee's hand can reach across the distance from the guard to the danger zone. Presence-sensing devices are typically installed on particle part-revolution power presses and other machinery.

Radiofrequency devices use a radio beam that is a part of the control circuit of the machine. Electronic circuitry that creates the field detects the presence of a human body. When an object is present in the danger zone, the capacitance field is interrupted, and the machine stops or will not operate. The sensitivity of the field is adjustable. Like the photoelectric device, this device is to be used only on machines that can be stopped before the worker can reach the danger area. This requires the machine to have a friction clutch or other reliable means for stopping.

Electromechanical devices use probes or contact bars that enter to a predetermined distance when the operator initiates the machine cycle. If an obstruction prevents it from descending its full predetermined distance, the control circuit does not actuate the machine cycle. These devices are commonly found on metals working presses.

See Table 4.1 for the advantages and limitations of presence-sensing devices.

Table 4.1. Pressure Sensing Devices – Advantages and Limitations

METHOD	SAFEGUARDING ACTION	ADVANTAGES	LIMITATIONS
Photoelectric	• Machine will not start cycling when the light field is interrupted • When the light field is broken by any part of operator's body during the cycling process, immediate machine braking is activated	• Can allow freer movement for operator • Simplicity of use • Used by multiple operators • Provide passerby protection • No adjustment required	• Does not protect against mechanical failure • Limited to machines that can be stopped
Radiofrequency	• Machine cycling will not start when capacitance field is interrupted • When the capacitance field is disturbed by any part of the operator's body during the the cycling process, immediate machine braking is activated	• Can allow freer movement for operator	• Does not protect against mechanical failure • Antennae sensitivity must be properly adjusted; this adjustment must be maintained properly • Limited to machines that can be stopped
Electro-mechanical	• Contact bar or or probe travels a predetermined distance between the operator and the danger area. Interruption of this movement prevents the starting of machine cycle	• Can allow access at the point of operation	• Contact bar or probe must be properly adjusted for each application; this adjustment must be maintained properly
Pullback	• As the machine begins to cycle, the operator's hands are pulled out of the danger area	• Eliminates the need for auxiliary barriers or other interference at the danger area	• Limits movement of operator • May obstruct work space around operator
Restraint	• Prevents the operator from reaching into the danger area	• Little risk of mechanical failure	• Adjustments must be made for specific operations and for each individual • Requires frequent inspections and regular maintenance • Requires close supervision of the operator's use of the equipment • Limits movement of operator • May obstruct work space
Safety trip controls	• Stops machine when controls tripped	• Simplicity of use	• All controls must be manually activated

METHOD	SAFEGUARDING ACTION	ADVANTAGES	LIMITATIONS
Pressure-sensitive body bar			• May be difficult to activate controls because of their location
safety triprod			• Only protects the operator
safety tripwire			• May require special fixtures to hold work • May require a machine brake
Two-hand control	• Concurrent use of both hands is required, preventing the operator from entering the danger area	• Operator's hands are at a pre-determined location • Operator's hands are free to pick up a new part after first half of cycle is completed	• Some two-handed controls can be rendered unsafe by holding with arm or blocking, thereby permitting one-hand operation • Requires a partial machine with a brake • Protects only the operator
Two-hand trip	• Concurrent use of two hands on separate controls prevents hands from being in danger area when machine cycle starts	• Operator's hands are away from danger area • Can be adapted to multiple operations • No obstructions to hand feeding • Does not require adjustment for each operation	• Operator may try to reach into danger area after tripping machine • Some trips can be rendered unsafe by holding with arm or blocking, thereby permitting one-hand operation • Protects only the operator • May require special fixtures
Gate	• Provides a barrier between danger area and operator or other personnel	• Can prevent reaching into or walking into the danger area	• May require frequent inspection and regular maintenance • May interfere with operator's ability to see the work

4.2.2 Pullback Devices

Pullback devices (see Figure 4.2) use a series of light-weight cables attached to the operator's hands, wrists, and/or arms to prevent operator's appendages from entering the point-of-operation during the power stroke. This type of device is primarily used on power presses and other machines with stroking action. When the slide/ram is up between cycles, the operator is allowed access to the point of operation. When the slide/ram begins to cycle by starting its descent, a mechanical linkage automatically assures withdrawal of the hands from the point of operation. To ensure their effectiveness, these devices must be properly fitted and adjusted for each worker.

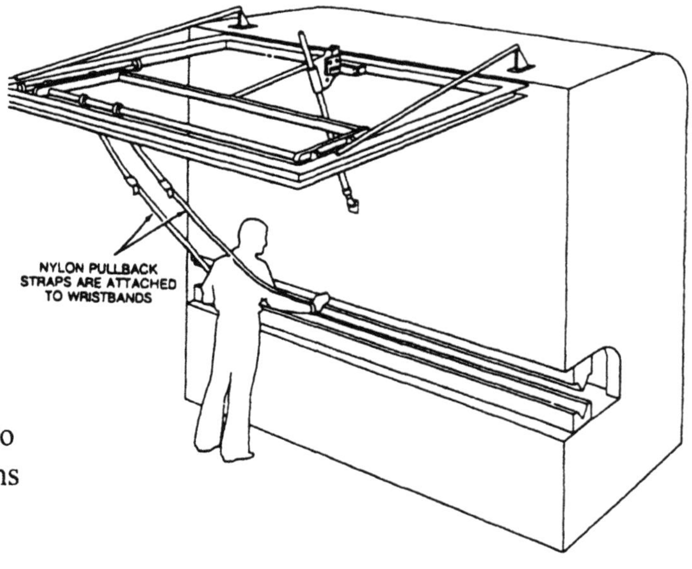

Figure 4.2. Example of a pullback device (from OSHA 3067).

See Table 4.1 for the advantages and limitations of pullback devices.

4.2.3 Restraints

Restraints (holdouts) also use cables or straps attached to the operator's hands at a fixed point to prevent hands from entering the point-of-operation. The cables, however, are attached to a fixed object instead of the machine. The cables or straps must be adjusted to allow the operator's hands to travel within a predetermined area, but by limiting the length of the cable, the operator's hands are prohibited from moving into the danger zone. No extending or retracting action is involved. Consequently, the operator cannot reach into the point-of-operation, even when the machine is not operating, necessitating the need for hand-feeding tools (see Figure 4.3) if the operation involves placing material into the danger area.

See Table 4.1 for the advantages and limitations of restraints.

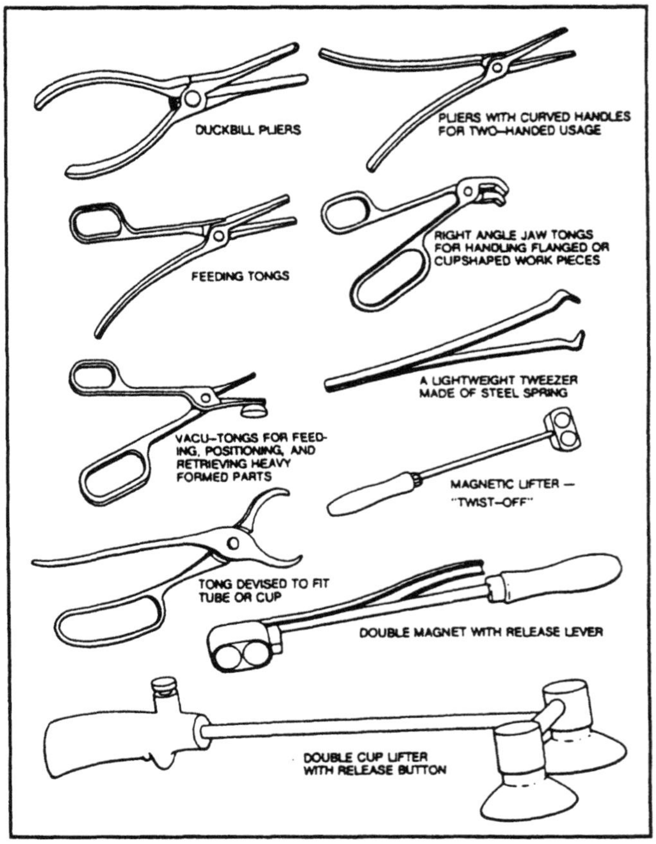

Figure 4.3. Examples of hand feed tools (from OSHA 3067).

4.2.4 Safety Trip Controls

Safety trip controls provide a quick means for deactivating the machine in an emergency situation. Safety trip controls include two-hand controls and two-hand trips. Safety trip controls allow the operator to quickly disengage the machine in emergency situations by applying pressure to a pressure-sensitive body bar. When depressed, the body bar deactivates the machine. If the operator trips, loses balance, or is drawn toward the machine, applying pressure to the bar stops the operation. The positioning of the bar, therefore, is critical. It must stop the machine before a part of the worker's body reaches the danger zone.

Two-hand controls (see Figure 4.4) require the operator to maintain a constant, concurrent hand pressure on two linked controls until the machine completes the dangerous part of the cycle. With the operator's hands on the controls, they are safely away from the point-of-operation. The controls must be designed so that they cannot be operated with one hand and another part of the body. Some two-hand controls have a time delay limit between activation of two controls. This prevents someone from pressing one control, waiting or doing something else, then pressing the second control with the hand off the first one.

Note: This kind of control requires a part-revolution clutch, brake, and a brake monitor if used on a power press.

Two-hand trips require the operator to press two buttons concurrently to start the machine. However, unlike two-hand controls, the operator's hands are free to perform other tasks after the machine is engaged. This device is usually used with machines equipped with full-revolution clutches. To provide adequate protection, the buttons are located so that the time required for the operator's hands to reach the danger zone is greater than the time required for the machine to complete the dangerous part of the cycle.

See Table 4.1 for the advantages and limitations of safety trip controls.

Note: To be effective, both two-hand controls and two-hand trips must be located so that the operator cannot use two hands or one hand and another part of his/her body to activate the machine.

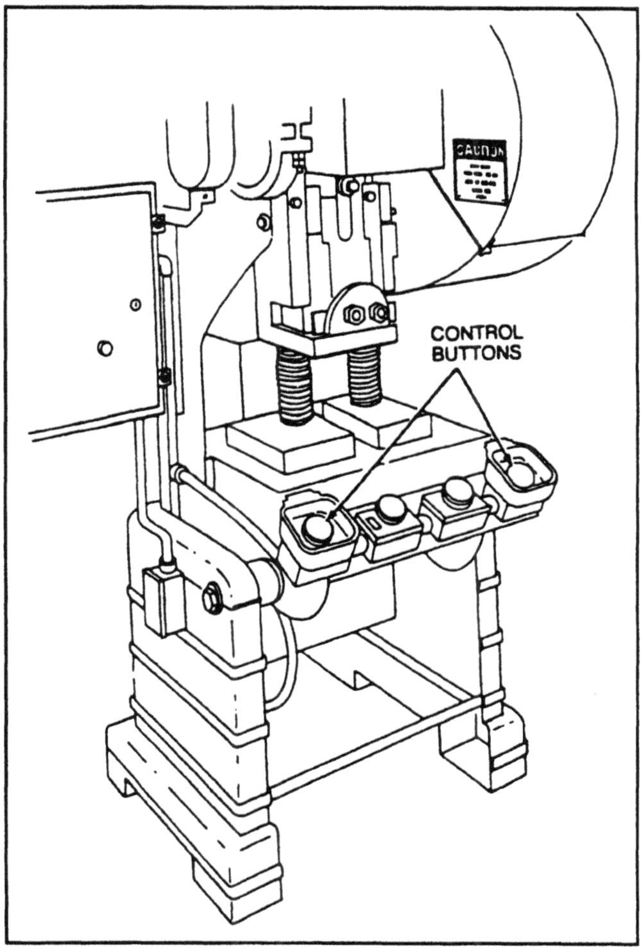

Figure 4.4. Example of a two-hand control device on a press (from OSHA 3067).

4.2.5 Gates

A *gate* is a movable barrier that protects the operator at the point of operation, before the machine cycle can be started. Gates can provide a high degree of protection to both the operator and other workers in the area, especially when interlocked to the machine cycle. Gates are, in many instances, designed to be operated with each machine cycle. When the gate is opened, the machine will not operate. However, gates that are not interlocked and serve only as a barrier necessitate the need for other precautions to protect employees when working inside the gate.

See Table 4.1 for the advantages and limitations of gates.

SUMMARY

Making the decision for safeguarding requirements for every different machine is not simple. But remember, one of the key concepts of machine guarding is that the further away from the point of operation workers can stay (and still effectively perform their work), the less their risk of machine-caused injury.

CITED REFERENCE

Puddicombe, R., A Guide to Using Safety Interlock Switches. *Occupational Hazards.* pp. 45-50, March, 1998.

CHAPTER 5

AUTOMATIC AND SEMIAUTOMATIC FEEDING AND EJECTING METHODS

***A recent OSHA citation for a machine safeguarding violation:*[4]**

"Doesn't Guard Saw; Fined After Severe Hand Injury"

Company: School District, Washington State

Business: School District

Why Inspected? Investigation into worker safety after a student suffered a severe hand injury using a saw in the shop area of the school—the agency determined workers were exposed to the same hazards.

Fine: $80,120

Reasons for fine:

Two willful citations for failure to:

- *Guard the blade of a 10-inch table saw in the shop area*
- *Ensure proper functioning of a 12-inch radial arm saw.*

Thirteen serious citations, including failure to:

- *Properly equip the table saw to protect operators from kickback and other malfunction hazards*
- *Equip a belt sander with a side guard*
- *Properly store oxygen cylinders*
- *Ensure the handle on a walk-in freezer functioned properly.*

---

[4] From *Safety Compliance Alert*, p. 4, July 23, 1998.

5.1 INTRODUCTION

One means for preventing the need to place body appendages such as fingers and hands into the point-of-operation is the use of *automatic* (see Figure 5.1) or *semiautomatic feed and ejection equipment*. Automatic devices require no action by the operator and semiautomatic devices only require the operator to perform a control function. In this chapter, we describe automatic and semiautomatic feeding and ejecting methods, and the advantages and limitations of each type.

Note: The key characteristic of automatic and semiautomatic feeding and ejecting methods is the elimination of the need for the operator to work in the point-of-operation.

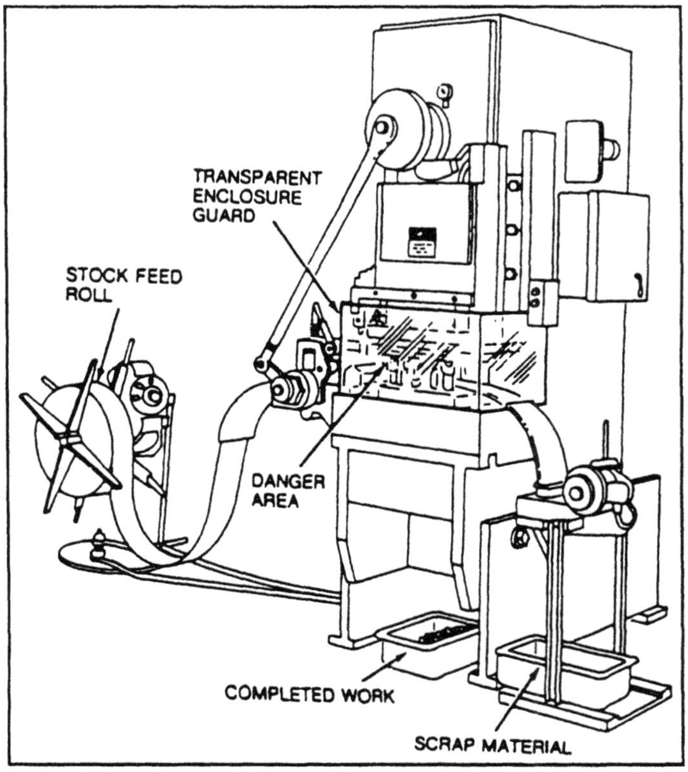

Figure 5.1. Example of a guard combined with an automatic feed (from OSHA 3067).

5.2 FEEDING AND INJECTION METHODS AND OPERATOR SAFETY

After a machine equipped with automatic or semiautomatic feeding and ejection systems is set up (ready for operation), no operator involvement is necessary. In other situations, operators can manually feed the stock with the assistance of a feeding mechanism. Properly designed ejection methods do not require any operator involvement after the machine starts to function.

However, keep in mind that some feeding and ejection methods may create hazards themselves. For instance, a robot may eliminate the need for an operator to be near the machine, but may create a new hazard itself by the movement of its arm.

Note: Using these feeding and ejection methods ***does not*** *eliminate the need for guards and devices. Guards and devices must be used whenever necessary to provide protection from, or exposure to hazards. Some systems may actually create other hazards—such as excessive noise from air ejection blow offs and/or flying objects.*

5.3 TYPES OF FEEDING AND EJECTING METHODS

Automatic feeds reduce the involvement by the operator, and sometimes do not require any effort (usually only monitoring) by the operator after the machine is set up and running. These use mechanical actions, roll stock systems, robots, forced air systems, gravimetric force, and other mechanical means to feed and move the parts from location to location.

With *semiautomatic feeding and ejection* (a power press, for example), the operator uses a mechanism to place the piece being processed in a chute, plunger, magazine, or other device for subsequent feeding into the point-of-operation. The operator does not need to reach into the danger area; the danger area is completely enclosed.

As we stated previously, though, automatic and semiautomatic feeding and ejecting methods do not eliminate the need for machine guards or devices around the point-of-operation. Other methods of safeguarding may also be required during maintenance operations.

We list the advantages and limitations of various automatic and semiautomatic feeding and ejection methods in Table 5.1.

Summary

Automatic and semiautomatic feeding and ejection devices are a perfect example of the advantage of keeping workers clear of the point of operation—the subject of Chapter 6. Of course, these methods cannot be applied to every machine and every operation, so we also discuss other methods of distancing the worker from dangerous machine positions.

Table 5.1. Feeding and Ejection Methods

Safeguarding Action	Advantages	Limitations
Automatic Feed		
• Stock is fed from rolls, indexed by machine mechanism, etc.	• Eliminates the need for operator involvement in the danger area	• Other guards are also required for operator protection-usually fixed barrier guards • Requires frequent maintenance • May not be adaptable to stock variation
Semiautomatic Feed		
• Stock is fed by chutes, movable dies, dial feed, plungers, or sliding bolster		
Automatic Ejection		
• Work pieces are ejected by air or mechanical means		• May create a hazard of blowing chips or debris • Size of stock limits the use of this method • Air ejection may present a noise hazard
Semiautomatic Ejection		
• Work pieces are ejected by mechanical means which are initiated by the operator	• Operator does not have to enter danger area to remove finished work	• Other guards are required for operator protection • May not be adaptable to stock variation
Robots		
• They perform work usually done by operator	• Operator does not have to enter danger area • Are suitable for operations where high stress factors are present, such as heat and noise	• Can create hazards themselves • Require maximum maintenance • Are suitable only to specific operations

CHAPTER 6

LOCATION AND DISTANCE AND SAFEGUARDING ACCESSORIES

Unsafe machinery is as dangerous to well-trained employees as untrained employees—even with excellent equipment—are to themselves.

— S. A. Brubaker, p. 37, 1997.

6.1 INTRODUCTION

Without a doubt, before attempting any safeguarding technique, good safety practice requires the performance of a thorough hazard analysis of each machine and particular situation. When considering safeguarding by *location and distance*, a thorough hazard analysis of each machine and particular situation is ***absolutely*** essential.

A variety of tools and methods may help to lower the hazard potential created by certain machines, even though they do not provide full or complete machine safeguarding. In this chapter, we discuss both safeguarding by location and distance, as well as miscellaneous safeguarding accessories.

6.2 SAFEGUARDING BY LOCATION AND DISTANCE

By analyzing the hazard, and strategically locating machine parts and/or the operator, effective safeguarding by location and distance can be accomplished.

You don't need to be a rocket scientist to understand that mechanical transmission equipment that is out of the reach of workers (and others) may not require safeguards. But again, we emphasize that careful analysis of each and every machine must be conducted to ensure location and distance will in fact protect people under normal circumstances.

For a part of a machine to be truly safeguarded by **location**, the dangerous moving parts of a machine must be positioned so that those parts are not accessible, or do not present a hazard to a worker during the normal operation of the machine. This may be accomplished by placing a machine so that the machine's hazardous parts are located away from operator workstations or other areas where employees walk or work. Positioning a machine with its power transmission apparatus against a wall and leaving all routine operations conducted on the other side of the machine is often a successful solution. Enclosure walls or fences can restrict access to machines. Enclosure screens, walls, or fences should be permanent, substantial, and at least 8 ft high. Another possible solution is to have dangerous parts located high enough to be out of the normal reach of any worker. Only trained personnel should have access to enclosed locations.

Feeding processes can be safeguarded by location, if a **distance** can be maintained to protect the worker's hands. The minimum distance required from a floor or walking surface to mechanical power transmission apparatus varies with different standards, although 7 or 8 ft is often used.

How is the proper distance calculated? Good question. OSHA provides the following formula to establish the safe distance.

$$D = 63\ in/s \times T_S \qquad (6.1)$$

where

- D is inches from the point-of-operation
- 63 in/s is an assumed hand speed
- T_s is the stopping time of the machine in seconds

The dimensions of the stock being worked on may provide adequate safety. For instance, if the stock is several feel long, and only one end of the stock is worked on, the operator may be able to hold the opposite end while the work is being performed and maintain a safe distance. Often, single-end punching machines can be operated in this way. However, depending upon the machine, protection might still be required for other personnel.

A combination of location and distance is also an effective means of safeguarding operators and others. For example, the remote positioning of the operator's control station provides safeguarding by location, which, in turn, also usually provides protection by distance. That is, when operator controls are at a safe distance from the machine, the operator is protected.

6.3 MISCELLANEOUS SAFEGUARDING ACCESSORIES

Even though they do not provide full or complete machine safeguarding, a variety of methods and tools may help to lower the hazard potential created by certain machines. Sound judgment is needed in their application and usage. We describe several examples of possible applications below.

6.3.1 Awareness Barriers/Signals

An *awareness barrier* does not provide physical protection, because it does not prevent access to the point-of-operation. Instead, it serves only to remind people of potential hazards by directing their attention to the fact that they are approaching the danger area. For an employee to enter the danger area, an overt act must take place—the employee must either reach or step over, under or through the barrier. Awareness barriers work two ways. They may consist of a guard rail, gate or similar barrier that prevents easy access or entry to the hazard area, or they may contact any part of the body of a person exposed to the point-of-operation to alert them that they are near or approaching a hazard. In conjunction with an awareness barrier, a posted warning should explain the hazard and the purpose for the barrier.

> *Note: Generally, awareness barriers are not considered adequate when continual exposure to the hazard exists.*

Awareness signals are audio and/or visual signals used with point-of-operation safeguards. The signals alert an operator or others that a hazard exists or is approaching.

6.3.2 Shields

Another miscellaneous aid or accessory, *Shields* may be used to provide protection from flying particles, or splashing cutting oils or coolants.

6.3.3 Special Devices or Tools

Special devices or *tools* that allow the operator's hands and arms to remain away from the point of operation can be used for placing objects in power presses or power brakes. Hand feed tools (see Figure 4.3) are special tools that allow an operator to place materials into the point-of-operation or remove them without inserting hands or fingers. On some machines, (table saws, jointers and belt sanders, for example), push blocks, push sticks and jigs have a similar use (usually for feeding stock into a saw blade)—they keep hands and fingers away from the point-of-operation. By doing this, they may provide a few inches of safety and prevent a severe injury.

> *Note: Holding tools should not be used **instead** of other machine safeguards; they are merely a supplement to the protection that other guards provide.*

6.3.4 Anti-Kickback Devices and Spreaders

Anti-kickback devices and *spreaders* help to prevent work from being thrown back at the operator, particularly with certain woodworking machines—circular and radial saws, for example. An anti-kickback device allows material to move freely into the saw blade, but pinches the material against the saw table when the material is pulled or pushed backward. Anti-kickback features, a similar device, protects radial saw operators during ripping operations.

6.3.5 Emergency Stop Devices

Emergency stop devices are required for situations when a worker might get caught in a machine or point-of-operation. An emergency stop device is an electrical, pneumatic, mechanical, or other control used to stop or deenergize a machine when an emergency occurs. An emergency control overrides all other controls, requiring a separate control to restart the machine. Emergency stop controls must be located so that operators or others have swift and easy access to the control. Usually larger than other machine controls so they are easy to locate and operate, stop controls must also be clearly labeled and painted red. They must require only momentary contact to activate. In machine operations where both hands are occupied when an emergency could occur, a foot operated emergency stop is called for (see Figure 6.1).

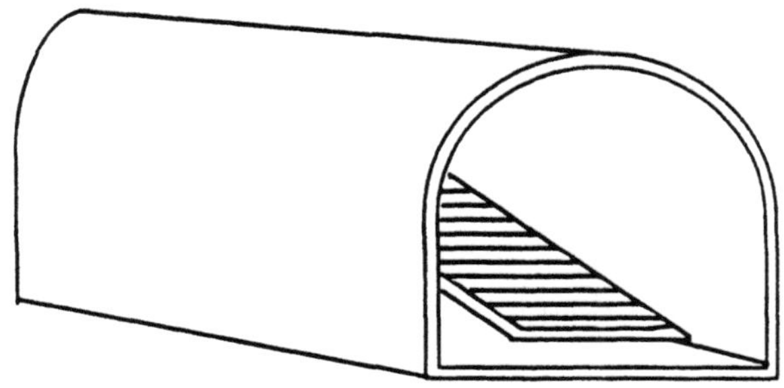

Figure 6.1. This foot control emergency stop is guarded on the top and sides.

> *Note: When an operator keeps a foot on a foot control between cycles, activating it at the wrong time is possible. To help prevent this dangerous action, foot controls should have a significant resistant force, and require an amount of movement consistent with intentional actions. Usually, human factors engineers factor in the correct amount of force required for minimum switch movement.*

> *Note: If multiple machine operators use a machine at the same time, each should have easy access to an emergency stop control. Emergency stop controls may be combined with braking devices to stop the machine quickly.*

6.3.6 Anti-Repeat Function

In machines that operate one cycle at a time, operators are at risk of injury whenever they become used to the rhythm of the machine. Normally, an operator inserts a part into the machine, then activates the machine, which operates on that part and stops. If, on occasion, the machine cycles more than once, the operator may have reached into the point-of-operation at the end of the first cycle, creating the possibility that an injury could occur. Wherever this situation is possible, the machine action must incorporate an *anti-repeat* device as a safeguard to prevent an inadvertent extra cycle.

6.3.7 Brakes

Certain presses have elevated components that may fall because of gravity. In these situations, a mechanical *brake* is usually built in to keep the components in position until a control cycles the machine. Some machines have brakes that can interrupt a machine at any point in its cycle.

6.3.8 Jog Controls

During troubleshooting, cleaning, setup operations, or routine maintenance, moving machine components to certain positions is often necessary. To accomplish this, an inching or *jogging control* is usually installed. Jog controls allow the machine to be turned on and off quickly in situations where the danger is in turning a machine on for too long and getting caught, or creating some other hazard. When activated, a jog control runs the machine; when deactivated, the control stops the machine without any additional action on the part of the operator.

6.3.9 Blocks and Stops

The ram on a press is a hazard to the operator both during operation, maintenance, and setup activities. To protect the worker during these latter operations, a die block or safety block is used to prop it open, to prevent the ram on the press from falling (see Figure 6.2).

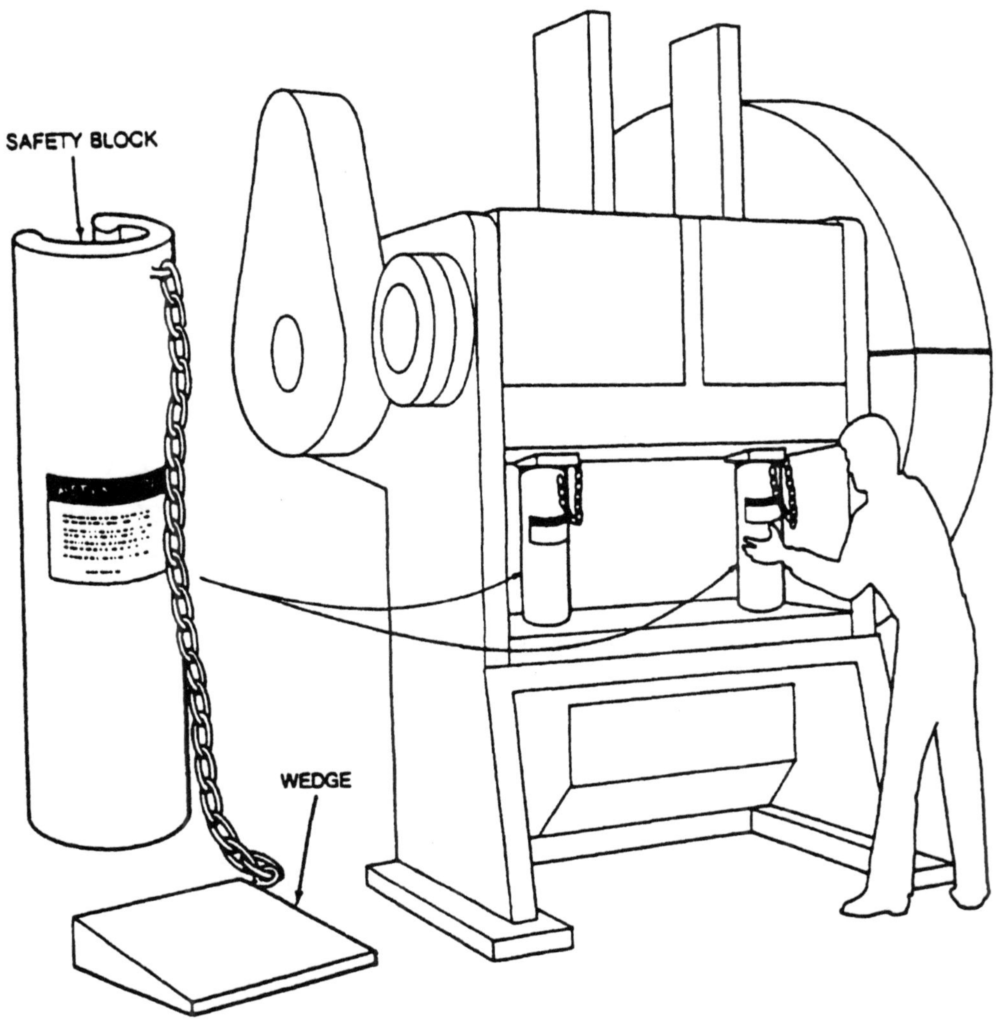

Figure 6.2. Example of a safety block for presses (from OSHA 3067).

6.2.10 Warnings

As with any other workplace hazard, one or more *warnings* are required on machines to communicate hazards that may be present. Exactly what type or kind of warnings should be posted? Good question. Figure 6.3 provides a few examples.

Figure 6.3 Example warning signs.

Warnings should state the hazards or danger, any limitation the device may have, and protective actions that operators must take. Effective warnings and instructions must have many characteristics based on good writing skills, knowledge of use environments, ergonomic principles, and other factors. Table 6.1[5] lists 15 important characteristics of effective warnings.

TABLE 6.1. CHARACTERISTICS OF EFFECTIVE WARNINGS

READABILITY: *The ability to read or receive a message.* Multiple languages, pictorials or symbolics, and Braille are all methods to ensure that a message is received.

UNDERSTANDING: *The ability to understand individual components of a message.* Some words are beyond the vocabulary of certain readers. Not all symbols are recognized or understood by every viewer. Multiple communications—by words, symbols, color, and shape, for example, allow higher levels of recognition.

COMPREHENSIBILITY: *The ability to understand the overall message.* Messages must be simply stated, require little technical or specialized knowledge, and be precise.

PRACTICALITY: *The ability to heed or comply with a warning.* Warnings must be interpreted in light of behavior that is normally expected, or given in a normal context for the warning.

EFFECTIVENESS: *The ability to prove a warning's effectiveness.* Valid and reliable test data establishes whether a warning does, in fact, communicate its message and is not just assumed to do so by its writer or designer.

BEHAVIOR MODIFICATION: *The ability to affect workers' actions.* An effective warning achieves the behavior desired by the warning, preventing unsafe or injury-causing acts that might otherwise occur.

COMPATIBILITY: *The ability to convey a message suitable for and consistent with expectations of individual applications.* Warnings should agree with local customs and practices, be consistent in similar situations (standardized), meet requirements of consensus and local standards, and be appropriate for a particular application situation.

CONSPICUOUS: *The ability to provide a reasonable certainty of perception, without search, and in a short time.* This characteristic includes size, color contrast, stimulus novelty, brightness level, and other characteristics.

DURABILITY: *The ability to resist environmental conditions.* Warning signage must remain effective for a reasonable lifespan, through and despite abrasion, wear, wetness, chemicals, sunlight, and other adverse conditions.

RELIABILITY: *The ability to function consistently.* Warnings must be present when needed. This property is particularly applicable to visual and audio warning devices that must act when a danger is present.

REINFORCEMENT: *The ability for repeated exposure to the warning.* Workers should be made aware of additional or more detailed data about a warning or its importance through training sessions, operating manuals, or other means. The goal is to influence the receiver's sensitivity toward the warning.

---

[5] Adapted from G. A. Peters, "15 Cardinal Principles to Ensure Effectiveness of Warning Systems," *Occupational Health and Safety*. pp. 76-79, May 1984.

DANGER SIGNAL: *The ability to attract worker attention.* Warnings should have attention-getting enhancements, such as underlined or boxed text, bright colors, signal words like *danger* or *warning*, special auditory tones, and so forth.

PLACEMENT: *The ability to maintain visual exposure.* Warnings must be located where they are likely to be seen or heard, in the location of the danger addressed; proximity in distance and time.

NOVELTY: *The ability to maintain effectiveness over time.* Use of attention-getting features like animation, voice synthesized messages, and color can ensure worker attention when activated.

TYPE: *The ability to address distinct levels of hazard communication.* Warnings fall under several different classifications of purpose or function, including (a) advisory, (b) explaining what to do, (c) reminder, and so forth.

Summary

All of the techniques, tools, warnings, and training that we use to keep workers from injury and accident, function, at least in part, to prevent worker complacency. When workers become too accustomed to their work, when they fall into thoughtless, automatic patterns, their risk of injury increases. The fictional production-line dreamer automatically going through the motions of the job is the last worker we want operating heavy machinery. Attention to the work at hand and attention to safe work practices protects the worker, and signage, tools, and safety devices function as reminders—to keep worker attention on the job.

Cited Reference

Brubaker, S. A., Optimum Machine Guarding. *Occupational Health and Safety*. pp. 37-40, December, 1997.

CHAPTER 7

ROBOTS IN THE WORKPLACE

Every machine, cell, system, or process has "points of hazard" or "areas of hazard." These include pinch points, web to roll nips, punching, shearing, hot surfaces, blades, flammable vapors, and protrusions. The locations where operators, maintenance personnel, other employees, and/or visitors are required to run, service, or pass by a machine, system, or process are considered "points or areas of operation."

A safety problem exists when a point or area of hazard coincides with a point or area of operation. Each hazard must be reviewed in relation to the task(s) of each person approaching or using the machines.

— J. Steinmann, p. 25, December, 1998.

7.1 INTRODUCTION[6]

Before we begin our discussion of robots in the workplace, and specifically, the safeguarding requirements involved with such advanced devices, we must again define a few key terms directly related to the subject area.

Obviously, the first term defined is *robot*—a reprogrammable, multifunctional, mechanical manipulator that typically employs one or more means of power: electromechanical, hydraulic, or pneumatic.

---

[6]*Much of the following information is derived from OSHA Instruction Pub 8-1.3, September 21, 1987 and OSHA's 3067 (1992 - Revised).*

Robots are used to replace human operators in performing unsafe, hazardous, highly repetitive, and unpleasant tasks. They are used chiefly to accomplish many different types of application functions—spray painting, spot-welding, and transfer and assembly tasks, for example. A robot performs its tasks in a physical area known as the robot operating work envelope. This work envelope is the area swept by all possible programmable robot movements (see Figure 7.1). This includes the area where work is performed by robot tooling.

A robot can have one or more arms, interconnected sets of links and powered joints. Arms are comprised of manipulators that support or move wrists and end-effectors. An *end-effector* is an accessory tool specifically designed for attachment to a robot wrist to enable the robot to perform its intended task. Examples of end-effectors include grippers, spot-weld guns, and spray paint guns. The ANSI R 15.06-1986 Standard defines an *industrial robot system* as that which includes industrial robots, end-effectors, and any equipment, devices, and sensors required for the entire robot system to perform its tasks.

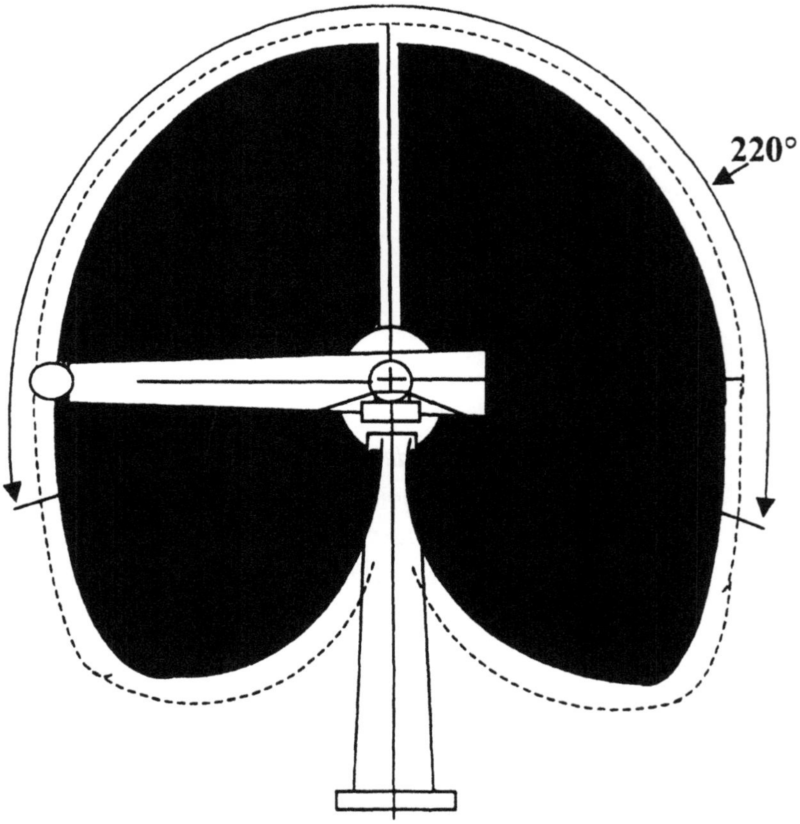

Figure 7.1. A robot's work envelope.

Most robots are set up for an operation by the *teach-and-repeat technique*. In this technique, a trained operator (programmer) typically uses a portable control device (commonly referred to as a *teach pendant*) to manually key a robot and its tasks. Program steps are of the up-down, left-right, in-out, and clockwise-counterclockwise variety. Robot speeds during these programming sessions must be slow. The ANSI Standard currently recommends that this slow speed should not exceed 10 in/sec (250 mm/sec).

The very nature of robotics systems operations has introduced a new type of employee into the industrial workplace, the corrective maintenance worker. This individual is normally present during all operations of a robotics system and is responsible for assuring continuing operation—adjusting speeds, correcting grips, and freeing jam-ups. The corrective maintenance worker may also be the trained programmer who guides a robot through the teach-and-repeat technique. This individual must be near the robot from time to time, which raises concerns about his or her safety and the safety of other workers who may also have access to the work envelope.

The age of robotics has brought about a recent development in manufacturing technology. Robotics has given rise to a manufacturing concept known as *Manufacturing Systems/Cells*, or *Cellular Manufacturing Systems*. These systems of integrated industrial machines, linked by a material handling system and operated by (controlled by) a programmable electronic system (computer) are capable of manufacturing discrete parts or assemblies. The safety of employees exposed to these systems is of first order importance.

> *Note: When a system must be safeguarded, logic dictates that a system hazard analysis will yield the parameters of the safeguarding system required.*

In this chapter, we discuss robot safety systems, guarding methods, control devices, installation, maintenance, programming, and training requirements.

7.2 SAFETY SYSTEMS FOR ROBOTS

The proper selection of an effective robotics safety system must be based on hazard analysis of the operation involving a particular robot. Among the factors under consideration in such an analysis are the task a robot is programmed to perform, the start-up and programming procedures, environmental conditions and location of the robot, requirements for corrective tasks to sustain normal operations, human errors, and possible robot malfunctions. According to the National Safety Council (1991), the principal hazards associated with robots are as follows:

- Being struck by a moving robot while inside the work envelope.
- Being trapped between a moving part of a robot and another machine, object, or surface.
- Being struck by a workpiece, tool, or other object dropped or ejected by a robot.

According to OSHA 3067, sources of robot hazards include:

- Human errors
- Control errors
- Unauthorized access
- Mechanical hazards
- Environmental hazards
- Electric, hydraulic, and pneumatic power sources.

Human errors can result in hazards both to personnel and equipment. Errors in programming, interfacing peripheral equipment, and connecting input/output sensors can all result in unpredicted movement or action by the robot, which can result in personnel injury or equipment breakage.

Human errors in judgment result frequently from incorrectly activating the teach pendant or control panel. The greatest human judgment error results from becoming so familiar with the robot's redundant motions that personnel are too trusting in assuming the nature of these motions, and place themselves in hazardous positions while programming or performing maintenance within the robot's work envelope.

Robots in the workplace are generally associated with the machine tools or process equipment. **Robots are machines**, and as such must be safeguarded in ways similar to those presented for any hazardous remotely-controlled machine.

7.3 SAFEGUARDING METHODS

Various techniques are available to prevent employee exposure to the hazards that can be imposed by robots. For example, the National Safety Council (1991) recommends several strategies for minimizing the hazards associated with robots. Those general strategies include:

- Ensure a glare-free, well-lighted robot site. The recommended light intensity is 50-100 footcandles.
- Keep the floors in and around the robot site carefully maintained, and clean and free of obstructions so that workers do not trip or slip into the work envelope.
- Keep the robot site free of associated hazards, such as blinding light from welding machines or vapors from a paint booth.
- Equip electrical and pneumatic components of the robot with fixed covers and guards.
- Clear the work envelope of all nonessential objects, and make sure all safeguards are in place before starting the robot.
- Apply lockout/tagout and proper test procedures before entering the work envelope.
- Remove and account for all tools and equipment used to maintain the robot before starting it.

In addition, various guarding methods can be employed to ensure safety around robots. In this section, we discuss many of the typical guarding methods.

7.3.1 Interlocked Barrier Guard

The most common technique is effected through installation of perimeter guarding with interlocked gates—a physical barrier around a robot work envelope that incorporates gates equipped with interlocks. A critical parameter relates to how the interlocks function. Specifically, the interlocks are designed so that all automatic operations of the robot and associated machinery will stop when any gate is opened. Restarting the operation requires closing the gate and reactivating a control switch located outside of the barrier. A typical practical barrier is an interlocked fence designed so that access through, over, under, or around the fence is not possible when the gate is closed.

7.3.2 Fixed Barrier Guard

A fixed barrier guard is a fence that requires tools for removal. Like the interlocked barrier guard, it prevents access through, over, under, or around the fence. Fixed barrier guards provide sufficient clearance for a worker between the guard and any robot reach, including parts held by an end-effector, to perform a specific task under controlled conditions.

7.3.3 Awareness Barrier Device

Low railings or suspended chains that define a safety perimeter function are *awareness barrier guards*. Intended to prevent inadvertent entry into the work envelope, they can be climbed over, crawled under, or stepped around. Such a device is acceptable only in situations where a hazard analysis indicates that the hazard is minimal, and interlocked or fixed guards are not feasible. Interlocked or fixed barrier guards provide a positive protection needed to prevent worker exposure to robotic systems hazards. Awareness barriers do not.

7.3.4 Presence Sensing Devices

The presence detectors that are most commonly used in robotics safety are pressure mats and light curtains. Floor mats (pressure sensitive mats) and light curtains (similar to arrays of photocells) can be used to detect a person stepping into a hazardous area near a robot. Proximity detectors operating on electrical capacitance, ultrasonics, radio frequency, laser, and television principles are currently undergoing reliability testing in research laboratories, because of recognized limitations in their ability to detect the presence of personnel. Although some of these devices are already available in the safety equipment marketplace, care must be used in their selection to ensure adequate safety and reliability. At this time, these proximity detectors are not recommended for such use, unless a specific analysis confirms their acceptability for the specific intended use.

Effective presence sensing devices stop all motion of the robot if any part of a worker's body enters the protected zone. They are designed to be fail-safe—a failure within the device will leave it unaffected, or convert it into a mode in which its failed state would not result in an accident. In some cases, this means deactivation of the robot. Factors considered in the selection of such devices include spatial limitations of the field, environmental conditions affecting the reliability of the field, and sensing field interference from robot operation.

7.3.5 Emergency Robot Braking

Dangerous robot movement is arrested by dynamic braking systems rather than simple power cut-off. Such brakes counteract the effects of robot arm inertia. Cutting off all power could create hazards such as a sudden dropping of a robot's arm or flinging of a workpiece.

7.3.6 Audible and Visible Warning Systems

Audible and visible warning systems are not acceptable safeguarding methods, but may be used to enhance the effectiveness of positive safeguards. Audible and visible signals must be easily recognizable as distinct warnings.

7.4 ROBOT CONTROL DEVICES

The following characteristics are essential for robot control devices:

- The main control panel is located outside the robot system work envelope and within sight of the robot.
- Readily accessible emergency stops (palm buttons, pull cords, etc.) are located in all needed zones. Clearly situated in easily located positions, the position identification must be a prominent part of personnel training. Emergency stops must override all other controls.
- The portable programming control device contains an emergency stop.
- Automatic stop capabilities are provided for abnormal robot component speeds and robot traverses beyond the operating envelope.
- All control devices are clearly marked and labeled as to device purpose. Actuating controls are designed to indicate the robot's operating status.

- Controls that initiate power or motion are constructed and guarded against accidental operation.
- Each robot is equipped with a separate circuit breaker that can be locked only in the "off" position.
- User-prompt displays are used to minimize human errors.
- The control system for a robot with a lengthy start-up time is designed to isolate power to components having mechanical motion, from the power required to energize the complete robot system.
- Control systems are selected and designed so that they prevent a robot from automatically restarting upon restoration of power after electrical power failure. The systems also prevent hazardous conditions in case of hydraulic, pneumatic, or vacuum loss or change.
- A robot system is designed to be moved manually on any of its axes without using the system drive power.
- All control systems meet OSHA 29 CFR 1910 Subpart S standards for electrical grounding, wiring, hazardous locations, and related requirements.

7.5 ROBOT INSTALLATION, MAINTENANCE, AND PROGRAMMING

Good installation, maintenance, and programming practices include the following:

- The robot is installed in accordance with the manufacturer's guidelines and applicable codes. Robots are compatible with environmental conditions.
- Power to the robot conforms to manufacturer's specifications.
- The robot is secured to prevent vibration, movement, and tip over.
- Installation must create no additional hazards, such as pinch points with fixed objects and robot components, or energized conductor contact with robot components.
- Signs and markings that indicate the zones of movement of the robot are displayed prominently on the robot itself, and if possible, on floors and walls.
- Stops are placed on the robot system's axes to limit its motions under rated load and maximum speed conditions.
- A lockout/tagout procedure is established and enforced for preventive maintenance or repair operations.
- The robot manufacturer's preventive maintenance schedule is followed rigorously.
- A periodic check of all safety-critical equipment and connections is established.
- Stored energy devices (such as springs and accumulators) are neutralized before robot servicing.
- Only programmers have access to the work envelope, and full control of the robot when it is in the teach mode.
- All robot motion initiated from a teach pendant used by a programmer located within the robot work envelope is subject to the current ANSI slow speed recommendation of 10 in/sec (250 mm/sec).

7.6 ACCIDENT PREVENTION TRAINING: ROBOTICS

Effective accident prevention programs include training. Some points to be considered in training programs include:

- Managers and supervisors in facilities that use robots must be trained in the working aspects of robots, so that they can set and enforce a robotics safety policy from an informed viewpoint.
- The employer ensures that his or her company has a written robotics safety policy that has been explained to all personnel who will be working with robots. This safety policy states by name which personnel are authorized to work with robots.
- Robot programming and maintenance operations are prohibited for persons other than those who have received adequate training in hazard recognition and the control of robots.
- Robot operators receive adequate training in hazard recognition and the control of robots, and in the proper operating procedure of the robot and associated equipment.
- Training is commensurate with a trainee's needs, and includes the safeguarding method(s) and the required safe work practices necessary for safe performance of the trainee's assigned job.
- If an authorized person must be within the work envelope while a robot is energized (for example, during a programming sequence), training is provided in the use of slow robot operation speeds and hazardous location avoidance until the work is completed. Such training also includes a review of emergency stops, and familiarization with the robot system's potentially hazardous energy sources.

Summary

Robots often are programmed to perform work that would be dangerous or unpleasant for human workers to perform, but they cannot replace human judgment. The most effective safeguard workers can use to protect themselves from injury while working with or near robots is to remember always that robots are utterly automatic, once they are programmed to a task.

Guards, safeguards, and protective devices as well as alert attention are essential means of worker protection against injury, wherever they are working. Another essential protective means—Lockout/Tagout—is the subject of Chapter 8.

Cited References

National Safety Council. *Robots.*, Data Sheet 1-717-85 Chicago: p. 1-3, 1991.

Steinmann, J., "Analysis Methods for New Work Cell Standards," *Occupational Health and Safety.* pp. 24-27, December, 1998.

CHAPTER 8

LOCKOUT/TAGOUT

A recent OSHA citation for a lockout/tagout violation:[*]

"Doesn't Protect Workers From Electric Shock"

Company: Company in Byron, IL

Business: Manufacturer

Why Inspected: Part of OSHA's national emphasis program on power presses

Fine: $166,950

Reasons for fine:

Five willful citations, including:

- *Failure to train employees in hazardous energy control procedures*
- *Failure to conduct periodic inspections on machines that could unexpectedly start or release stored energy and cause injury*
- *No lockout/tagout program*
- *Failure to lock out hazardous machines.*

Fourteen serious citations, including:

- *No guards on power presses*
- *No electrical safety training*
- *Lack of written electrical safety procedures*
- *Blocked fire doors.*

---

[*] *From* Safety Compliance Alert, *p. 4, February 19, 1998.*

8.1 INTRODUCTION

One of the most important and effective safeguarding approaches in use today is the *lockout/ tagout program*. Lockout/tagout is a method specifically designed to protect against the unexpected startup of a machine that the operator believes is turned off. Why is this so critically important? OSHA statistics show that approximately 6% of all workplace fatalities are the result of the unexpected activation of machines while they are being serviced, cleaned, or otherwise maintained (Caruey, 1991).

In this chapter, we discuss OSHA's 29 CFR 1910.147 Standard: Lockout/Tagout. We know this is one of the most important chapters in this text, because we know adherence to the requirements of the standard prevents damaging or fatal events from occurring. Caruey (1991) provides us with examples:

- an employee is cleaning the guarded side of an industrial saw when she is pulled into it and killed;
- an employee is cleaning spilled lubricant from under a shear when another employee activates the machine and decapitates him; and
- an employee is cleaning paper from a waste crusher when he falls into it and is crushed.

8.2 MACHINERY MAINTENANCE AND REPAIR

Not only do well-designed and manufactured machines provide the worker, the employer, and the owner safe and satisfactory service in performing their intended functions, but good maintenance and repair procedures also contribute significantly to the safety of the maintenance crew, the material condition of the machine, and the safety and well being of machine operators.

The variety and complexity of machines to be serviced, the hazards associated with their power sources, the special dangers that may be present during machine breakdown, and the severe time constraints often placed on maintenance personnel, all make safe maintenance and repair work difficult. Lockout/Tagout makes it substantially less dangerous.

If possible, machine design should permit routine lubrication and adjustment without removal of safeguards. But when safeguards must be removed so the machine may be serviced, the lockout procedure 29 CFR 1910.147 must be followed rigorously. The maintenance and repair crew must ***never*** fail to replace the guards before their job is finished and they release the machine from lockout.

If machine parts must be oiled while the machine is running, special safeguarding equipment may be needed solely to protect the oiler from exposure to hazardous moving parts. Maintenance personnel must know which machines can be serviced while running—and which cannot. "If in doubt, lock it out" is the rule. Obviously, the danger of accident or injury is reduced by shutting off and locking out all sources of energy.

In situations where the maintenance or repair worker would necessarily be exposed to electrical elements or hazardous moving machine parts in the performance of the job, ***all*** power sources must be shut off and locked-out before work begins. Warning signs or tags are inadequate insurance against the untimely energizing of mechanical equipment.

Thus, one of the first procedures for the maintenance person is to disconnect and lock out the machine from all of its power sources, whether the source is electrical, mechanical, pneumatic, hydraulic, or a combination of these. Energy accumulation devices must be "bled down."

In performing a lockout, the worker places a padlock through a gate covering the activating mechanism, or applies it in some other manner to prevent the machine from being turned on until the lock is removed. These locks usually carry a label that gives the name, department, and telephone extension of the person who put it on.

Note: OSHA requires that individual padlocks be employee identifiable—by whatever means chosen by the employer.

It may also carry a message such as the following: "*This lock is to be removed only by* _______________." The name in the blank is the person who applied the lock (see Figure 8.1).

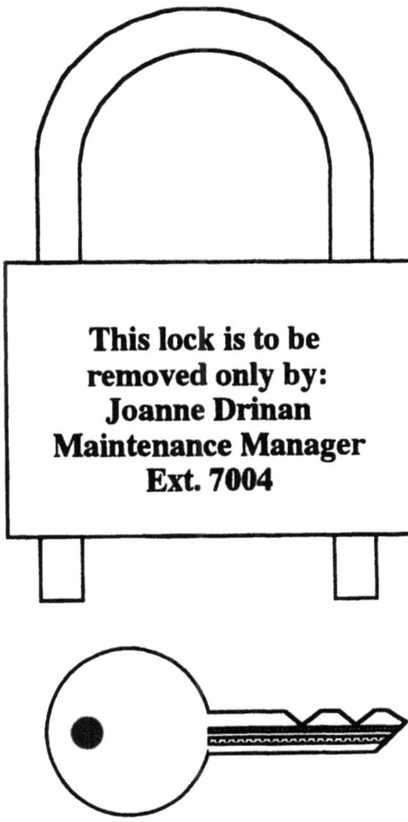

Figure 8.1. Lockout System.

In tagging out a system, the procedure is exactly like a lockout system, except a tag is substituted for the lock. Tags should be used only in cases where a lock is not feasible. Figure 8.2 shows an example of a tag, tag and lock (sometimes they are used together), and various other lockout/tagout devices presently available for use.

Unexpected energizing of any electrical equipment that can be started by automatic or manual remote control may cause electric shock or other serious injuries to the machine operator, the maintenance worker, or others operating adjacent machines controlled by the same circuit. For this reason, when maintenance personnel must repair electrically-powered equipment, they should open the circuit at the switch box and padlock the switch (lock it out) in the "off" position. This switch should be tagged with a description of the work being done, the name of the maintenance person, and the department involved. When more than one worker is engaged in the servicing/maintenance function, a typical lockout hasp to which each may affix a personal lock is used (see Figure 8.2).

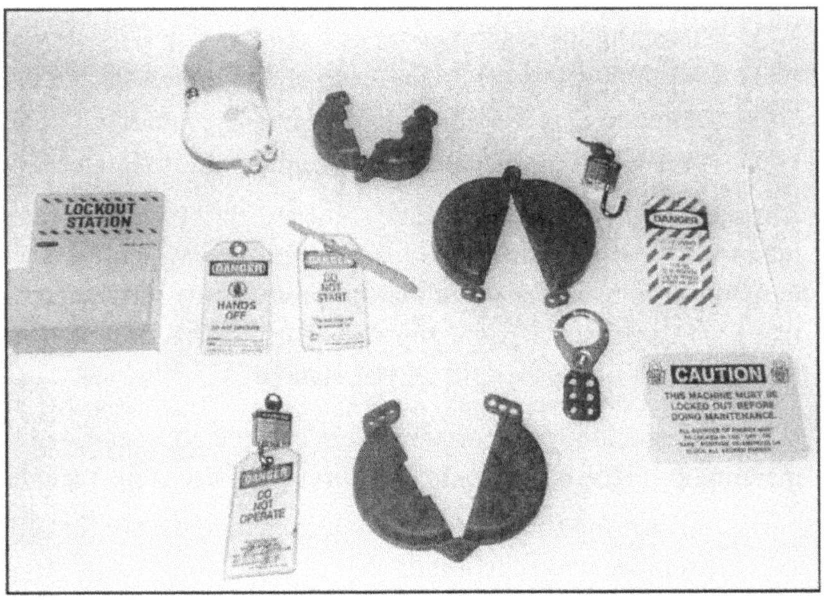

Figure 8.2. Lockout/Tagout Accessories.

In addition to standard lockout procedures, sometimes a safety block can be used on such machines as mechanical power presses. The safety block prevents the ram from coming down under its own weight.

In pneumatic and hydraulic systems, a lockout value is used to deenergize the machine. For example, the lever-operated air valve used during repair or shutdown to keep a pneumatic-powered machine or its components from operating can be locked open or shut. Before the valve can be opened, everyone working on the machine must use his or her own key to release the lockout. A sliding-sleeve valve exhausts line pressure at the same time it cuts off the air supply. Valves used to lock out pneumatic or hydraulic-powered machines should be designed to accept locks or lockout adapters, and should be capable of "bleeding off" pressure residues that could cause any part of the machine to move.

In shops where several workers might be working on the same machine, multiple lockout devices accommodating several padlocks are used. The machine cannot be reactivated until each person removes his or her lock. As a matter of general policy, lockout control is gained by the procedure of issuing personal padlocks to each repair or maintenance worker; no one but that person can remove the padlock. Each worker thus controls the power systems.

Whenever machines or equipment are serviced, hazards are encountered by the employees performing the servicing or maintenance that are unique to the repair of maintenance procedures being conducted. These hazards may exist because of failure of the employees doing the servicing or maintenance to stop the machine they are working on. Even if the machine has been stopped, the machine can still be hazardous because of the possibility that the machine could be re-energized or restarted.

To prevent these hazards, each machine or piece of equipment should be safeguarded during the conduct of servicing or maintenance by:

1. notifying all affected employees (usually machine or equipment operators or users) that the machine or equipment must be shut down to perform some maintenance or servicing;
2. stopping the machine;
3. isolating the machine or piece of equipment from its energy source;
4. relieving any stored or residual energy; and
5. verifying that the machine or equipment is isolated from the energy source.

Although this is the general rule, some exceptions exist—for example, when the servicing or maintenance is not hazardous for an employee, when the servicing which is conducted is minor in nature (done as an integral part of production) and the employer utilizes alternative safeguards that provide effective protection as is required by 29 CFR 1910.212 (General Requirements for Machines) or other specific OSHA standards.

When the servicing or maintenance is completed, specific steps must be taken to return the machine or piece of equipment to service. These steps include:

1. inspecting the machine or equipment to ensure that all guards and other safety devices are in place and functional,
2. checking the area to ensure that energization and start up of the machine or equipment will not endanger employees,

3. removing the lockout devices,
4. reenergizing the machine or equipment, and
5. notifying affected employees that the machine or equipment may be returned to service.

The lockout steps described above are only a part of the total energy control training that must exist in the workplace. The employee should also have written procedures for all machines and equipment. Employees must be trained in their duties and responsibilities under the energy control program, and periodic inspections must be conducted to maintain the effectiveness of the program.

Because (as we stated above) lockout/tagout is vital to ensuring a viable machine safeguarding program we have included more detailed guidance in the following sections, including a sample written program. Remember—as with almost all other OSHA-required safety programs—the program must be in writing.

8.3 OSHA's REQUIREMENTS FOR LOCKOUT/TAGOUT[8]

Employers are required to develop, document, and utilize an energy control procedures program to control potentially hazardous energy.

The energy control procedures must:

- *specifically outline the scope, purpose, authorization, rules, and techniques to be utilized for the control of hazardous energy; and*
- *the means to enforce compliance including, but not limited to, the following:*
  - *a specific statement of the intended us of the procedure;*
  - *specific procedural steps for shutting down, isolating, blocking, and securing machines and equipment to control hazardous energy;*
  - *specific procedural steps for the placement, removal, and transfer of lockout or tagout devices, and the responsibility for them;*
  - *specific requirements for testing a machine or equipment to determine and verify the effectiveness of lockout devices, tagout devices, and other energy control measures.*

— 29 CFR 1910.147

One of the major requirements in OSHA's Machine Guarding Standard is to ensure that the hazards within, on, or near a machine have been removed or isolated before maintenance or repair work is started. Removing any hazard is always the best way to protect employees; however, removing all hazards is, in many cases, impossible. Thus OSHA requires the control of hazardous energy using isolation, blanking or blinding, disconnection, and/or lockout/tagout procedures.

In our experience, many workers mistake the results of shutting down a machine as the same thing as deenergizing it. Nothing could be further from the true situation (in most cases). Many other dangers are inherent to machines, dangers that make maintenance or repair work on them hazardous. For example, if the machine has electrical devices and circuitry inside (and just about

[8]*From Spellman & Whiting,* Safety Engineering: Concepts & Applications. *Rockville, Maryland: Government Institutes, Inc., 1999.*

all of them do), an electrocution hazard may exist. If hazardous chemicals are fed to the machine, the potential for a hazard exposure is present. Many machines have hydraulic or pneumatic power sources—which are obviously energy sources—sources that must be isolated before the machine is worked on.

Simply stated, to ensure that the machine (any machine) is indeed safe, any and all sources of hazardous energy must be first isolated before maintenance or repair work is accomplished. The primary method used to accomplish this is through lockout/tagout procedures. In the following sections, we first define the key terms associated with lockout/tagout, then present a sample of a written lockout/tagout program. This written lockout/tagout program has a huge advantage over many other such programs: it has been used in the real world. It has been tested—it works.

8.3.1 Lockout/Tagout Key Definitions

Affected employee: An employee whose job requires him/her to operate or use a machine or equipment on which servicing or maintenance is being performed under lockout or tagout, or whose job requires him/her to work in an area where such servicing or maintenance is being performed.

Authorized employee: A person who locks out or tags out machines or equipment to perform servicing or maintenance on that machine or equipment. An affected employee becomes an authorized employee when that employee's duties include performing servicing or maintenance covered under the Organization's lockout/tagout program.

Capable of being locked out: An energy isolating device is capable of being locked out if it has a hasp or other means of attachment to which (or through which) a lock can be affixed, or it has a locking mechanism built into it. Other energy isolating devices are capable of being locked out if lockout can be achieved without the need to dismantle, rebuild, or replace the energy isolating device, or permanently alter its energy control capability.

Note: Whenever replacement or major repair, renovation, or modification of a machine or equipment is performed, and whenever new machines or equipment are installed, energy isolating devices for such machines or equipment shall be designed to accept a lockout device.

Energized: Connected to an energy source or containing residual or stored energy.

Energy isolating device: A mechanical device that physically prevents the transmission or release of energy, including (but not limited to) the following: A manually operated electrical circuit breaker, a disconnect switch, a manually operated switch by which the conductors of a circuit can be disconnected from all ungrounded supply conductors, and in addition, in which no pole can be operated independently; a line valve; a block; and any similar device used to block or isolate energy. Push buttons, selector switches, and other control circuit type devices are not energy isolating devices.

Energy source: Any source of electrical, mechanical, hydraulic, pneumatic, chemical, thermal, or other energy.

Hot tap: A procedure used in repair, maintenance, and services activities, which involves welding on a piece of equipment (pipelines, vessels, or tanks) under pressure, to install connections or appurtenances. Commonly used to replace or add sections of pipeline without the interruption of service for air, gas, water, steam, and petrochemical distribution systems.

Lockout: The placement of a lockout device on an energy isolating device, in accordance with an established procedure, ensuring that the energy isolating device and the equipment being controlled cannot be operated until the lockout device is removed.

Lockout device: A device that utilizes a positive means (such as a lock, either key or combination type) to hold an energy isolating device in the safe position and prevent the energizing of a machine or equipment. Included are blank flanges and bolted slip blinds.

Normal production operation: The utilization of a machine or equipment to perform its intended production function.

Selecting and/or maintenance: Workplace activities such as constructing, installing, setting up, adjusting, inspecting, modifying, maintaining, and/or servicing machines or equipment. These activities include lubrication, cleaning, or unjamming of machines or equipment, and making adjustments or tool changes, where the employee may be exposed to the unexpected energization or startup of the equipment or release of hazardous energy.

Setting up: Any work performed to prepare a machine or equipment to perform its normal production operation.

Tagout: The placement of a tagout device on an energy isolating device, in accordance with an established procedure, to indicate that the energy isolating device and the equipment being controlled may not be operated until the tagout device is removed.

Tagout device: A prominent warning device, such as a tag and a means of attachment, which can be securely fastened to an energy isolating device in accordance with an established procedure, to indicate that the energy isolating device and the equipment being controlled may not be operated until the tagout device is removed.

8.3.2 Lockout/Tagout Program (sample)

I. Purpose

This program has been developed to ensure protection of employees and to maintain compliance with OSHA Standard 1910.147, Control of Hazardous Energy.

These instructions establish the minimum requirements for the lockout or tagout of energy isolating devices whenever maintenance or servicing is done on machines or equipment. It shall be used to ensure that the machine or equipment is stopped, isolated from all potentially hazardous energy sources, and locked out before employees perform any servicing or maintenance. These procedures apply:

- whenever an employee has to remove or bypass a guard or other safety device;
- where such servicing results in the employee placing all or part of their body in a danger zone;
- where the unexpected energizing or start-up of the machine or equipment or release of stored energy could cause injury or death.

II. Authorized Employees

Authorized Organization employees shall be trained in lockout/ tagout procedures. Retraining will be done whenever an authorized employee's job responsibility changes or when the periodic inspection identifies procedural deficiencies. All **affected** (and other) employees whose work operation are or may be in the area shall be instructed in the purpose and use of lockout/tagout procedure (see Appendix A for list of **authorized** employees).

III. Application and Compliance

This procedure applies to all Organization employees and shall be enforced. All employees must signify that they have read and understand each part of this procedure by signing the training record (see Section IX of this procedure).

All employees are required to comply with the restrictions and limitations imposed upon them during the use of lockout or tagout. The employee, upon observing a machine which is locked or tagged out for servicing **shall not** attempt to start or use that machine.

IV. Tags

Tags are used to prevent the unexpected energization of equipment, and generally are restricted to those controls which cannot be locked or safeguarded by any other method.

- When the energy isolating device(s) cannot be locked out, use a tag to prevent inadvertent actuation.
- Tags will be attached to equipment or machinery, at the control panel, and at other points of operation.
- Tags must bear the words "DANGER," and "DO NOT OPERATE," or "DO NOT USE" printed on both sides of the tag.
- Tags shall bear the name, department, date, and telephone number of the employee (or department) performing the work.

V. Locks

Locks are employee-identifiable, key-operated padlocks. Lockout devices are those openings in equipment control handles or switches that can accept a padlock. Multiple lockout devices are those devices which have spaces for the application of more than one lock. This type of device is locked into the lockout opening, and subsequent locks prevent removal of the device.

Authorized employees will be assigned their own individual lock with one unique key. A master key will generally not be available except under very unusual circumstances. If a master key is necessary; it will be under the strict control of the work center supervisor. Locks will not be removed by anyone other than the authorized employee who placed it on the control, unless the procedure described in the "Special Condition" section of this program is followed. Employees working on the same machine will each use their own lock. When work continues from one shift to the next, employees leaving must remove their locks and employees beginning work must apply their own locks.

VI. Energy Analysis

You should include a statement similar to this sample in your own Lockout/Tagout Program.

> *In 19___, Organization's Safety Division surveyed each Organization facility and identified all machines or equipment covered by this program. All energy sources were identified, along with their methods of control. Each time a new machine is installed, the Organization's Safety Division is responsible for performing an energy analysis. This energy analysis procedure is on-going and must be updated as required.*

A. Equipment Energy Analysis Survey

Anytime a new machine or new configuration to an existing machine is installed, Organization's Safety Director conducted a survey of all Organization's properties and made the following findings:

(List all equipment that must be locked/taggedout here:

Equipment ______________ Equipment ______________

Equipment ______________ Etc. ____________________

—and various equipment/machinery identified in Organization's Maintenance Management System (preventive maintenance program), which have equipment-specific lockout/tagout procedures (see Appendix B). Organization is not required to document the equipment-specific procedure for each particular machine or equipment, because these machines and/or equipment:

- were found to have no potential for stored or residual energy, or re-accumulation of stored energy after shut down which could endanger employees;
- the machine or equipment has a single energy source which can be readily identified and isolated;
- the isolation and locking out of that energy source will completely de-energize and deactivate the machine or equipment;
- the machine or equipment is isolated from the energy source and locked out during servicing or maintenance;
- a single lockout device will achieve a locked-out condition;
- the lockout device is under the exclusive control of the authorized employee performing the servicing or maintenance;

- the service or maintenance does not create hazards for other employees; and
- Organization has had no accidents involving the unexpected activation or re-energization of the machine or equipment during servicing or maintenance.

For those machines and/or equipment that do not require equipment specific lockout/tagout procedures, the lockout/tagout procedure in Section VII is to be used at Organization.

VII. Lockout/Tagout Procedure

Lockout/tagout procedures for Organization equipment other than equipment specific lockout procedures for the Incinerators and other covered equipment (see Appendix B) are:

1. Notify appropriate operations and maintenance supervisors of lockout/tagout.
2. Place the main switch, valve, control, or operating lever in the "off", "closed", or "safe" position.
3. **CHECK** and **TEST** to **MAKE CERTAIN** that the proper controls have been identified and deactivated.
4. Place a lock to secure the disconnection whenever possible. If a lock cannot be used on electrical equipment, an electrician shall remove fuses or disconnect circuit.
5. If a system cannot be locked out with a lock, attach a HOLD-OFF, DO NOT ENERGIZE, or other such tag to the switch, valve, or lever. If the Organization work center does not use employee identifiable locks, a lock and tag must be used together.
6. When auxiliary equipment or machine controls are powered by separate supply sources, such equipment or controls shall also be locked or tagged to prevent any hazard that may be caused by operating the equipment or exposure to live circuits.
7. When equipment uses pneumatic or hydraulic power, pressure in lines or accumulators shall be checked. Using whatever safe means possible, this pressure shall be disconnected or pressure lines disconnected.
8. When stored energy is a factor as a result of position, spring tension, or counterweighting, the equipment shall be placed in the bottom or closed position, or it shall be blocked to prevent movement.
9. When the work involves more than one person, additional employees shall attach their locks and tags as they report.
10. When outside contractors are involved, the equipment shall be locked out and tagged in accordance with this procedure by the Project Manager supervising the work. Only in emergency cases is equipment to be shut down by other than an Organization representative.
11. When the servicing or maintenance is completed and the machine or equipment is ready to return to normal operating condition, the following steps will be used:
    a. Check the machine or equipment and the immediate area around the machine to ensure that tools, materials, and other nonessential items have been removed and that the machine or equipment components are operationally intact. Ensure that all guards have been replaced.
    b. Check the work area to ensure that all employees have been safely positioned or removed from the area.

c. Verify that the controls are in neutral.
d. Remove the lockout devices and re-energize the machine or equipment. Each lockout/tagout device will be removed from each energy isolating device by the employee who applied the lock.
e. Notify affected employees that the servicing or maintenance is completed, and that the machine or equipment is again ready for use.

VIII. Periodic Inspections

A periodic inspection of the energy control procedure(s) will be conducted at least quarterly to ensure that the requirements of the program and the standard are being followed to ensure full employee protection.

The periodic inspection will be performed by the work center supervisor and/or Safety Division personnel.

The inspection will be conducted to identify any program inadequacies that need correcting. The inspection will review the employee's responsibilities under the procedure being inspected.

Organization will provide **certification** that the inspection of a lockout/tagout has been performed. The certification will identify the machine on which the lockout/tagout is being utilized, the date of the inspection, names of employees involved, and the name of the individual performing the inspection. These entries will be made on the work center quarterly inspection form and the Periodic Inspection Worksheet shown below. These forms on to be kept on file by the Safety Division for one year.

Periodic Inspection Worksheet

(Certification Sheet)

Lockout/tagout program inspection for Organization work centers.

Name of authorized employee: ____________________________

Department	Machinery	Type of Power	Method of Control
___________	___________	Electrical	_______________
		Mechanical	_______________
		Hydraulic	_______________
		Pneumatic	_______________
		Chemical	_______________

Did the energy control procedures identify all hazards associated with this machine?

____

If not, describe deficiencies.

____

____

Is retraining required? ____

Has retraining been done? (Date) ____________________________________

______________________________ ______________________________

Signature of Inspector Date

IX. Training

Training will be provided by Organization's Safety Division, to ensure that the lockout/tagout program's purpose and function are understood by all employees, and that the knowledge and skills required for safe application, usage, and removal of energy controls are acquired by employees. The training shall include:

- Each **AUTHORIZED (person who actually locks/tags out)** employee shall receive training in the recognition of hazardous energy sources, the type and magnitude of the energy available in the workplace, and the methods and means necessary for energy isolation and control.
- Each **AFFECTED (person affected by the lockout/tagout)** employee shall be instructed in the purpose and use of the energy control procedure.
- All **OTHER** employees whose work operations are or may be in areas where energy control procedures may be used, shall be instructed about the procedure, and about the prohibitions relating to attempts to restart or re-energize machines or equipment which are locked out or tagged out.

When tagout is used, employees shall receive additional training on the limitations of tags.

A training record (see below) will be used to record each employee's training.

TRAINING RECORD
(sample)

TRAINING ROSTER

LOCKOUT/TAGOUT PROGRAM

DATE: ____________________

In accordance with the recording and training requirements of the Lockout/Tagout Program, OSHA 29 CFR 1910.147, I have received Safety training on Lockout/Tagout requirements and procedures. I have agreed to verify my understanding and training of 1910.147 by signing and dating this form. This training meets requirements of 29 CFR 1910.147.

Name	Work Center
____________________	____________________

X. Special Conditions

LOCKOUT/TAGOUT REMOVAL WHEN AUTHORIZED EMPLOYEE IS ABSENT

When the authorized employee who applied the lockout or tagout device is not available to remove it, that device may be removed under the direction of the supervisor, provided that specific procedures and training for such removal have been developed, documented and incorporated into the lockout/tagout program. Specific procedures include the following elements:

1. Verify that the authorized employee who applied the device is not at the facility;
2. Make all reasonable efforts to contact the authorized employee to inform him/her that his/her lockout or tagout device has been removed;
3. Ensure that the authorized employee has this knowledge before he/she resumes work at that facility; and
4. Complete the "Lockout/Tagout Removal When Authorized Employee is Absent" form—the form shall be kept by the work center supervisor.

LOCKOUT/TAGOUT REMOVAL
WHEN AUTHORIZED EMPLOYEE IS ABSENT

Department ____________ Machine/Equipment ____________ Lock Id # ____________

Authorized Employee ____________ Type of Work Being Done ____________

Verification that employee is not on site: (Yes) ___ (No) ___

Note: ____

____

Verification that employee has been informed before coming back to work that his/her lockout/tagout device has been removed. (Yes) _______ (No) ______

Note: ____

____

____________________ ____________________

Supervisor Date

XI. Methods of Informing Outside Contractors of Procedures

Whenever outside servicing personnel are to be engaged in lockout/tagout activities covered by this program, Organization **AND** the outside employer **WILL INFORM EACH OTHER** of their respective lockout or tagout procedures.

Organization employees are to be trained to understand and comply with the restrictions and prohibitions of the outside contractor's lockout/tagout energy control program.

The Safety Division will provide to the Organization work center supervisor a contractor briefing form (shown below) that will inform the contractor on precautionary measures involved in performing lockout/tagout on the site.

RECORD OF NOTIFICATION

ORGANIZATION AND OUTSIDE CONTRACTOR SAFETY ARRIVAL CONFERENCE

In accordance with the recordkeeping and training requirements under OSHA 1910.119 and other applicable standards, I have received a safety brief from Organization Safety Division/Plant personnel covering Organization's Lockout/Tagout Program. I further understand that Organization expects all outside contractors, including sub-contractors, to perform construction activities under OSHA required guidelines and Organization procedures. It is understood that all information regarding the Hazards, Safety Procedures, and Lockout/Tagout Procedures shall be disseminated to all persons, sub-contractors and agents employed either directly or indirectly by us. Regarding Lockout/Tagout procedures, Organization and the contractor must discuss these operations and work in a coordinated effort so as to prevent injury to anyone.

Signature	Printed Name	Organization Name
________________	______________	____________________________
________________	______________	____________________________
________________	______________	____________________________

APPENDIX A

LIST OF AUTHORIZED EMPLOYEES

APPENDIX B

EQUIPMENT-SPECIFIC LOCKOUT/TAGOUT PROCEDURES

SPECIFIED EQUIPMENT

Equipment ______________ Equipment ______________

Equipment ______________ Equipment ______________

Summary

Knowledge of machine guards and their use requires training, as does effective use of lockout/tagout, or any other workplace safe work practice or on-the-job operation. Training—the critical component for worker safety—is the subject of Chapter 9.

Cited References

Caruey, A., "Lock Out the Chance for Injury," *Safety and Health.* Vol. 143, 5, p. 46. May, 1991.

The Office of the Federal Register, *Code of Federal Regulations Title 29 Pars 1900-1910 (.147).* Office of Federal Registers, Washington, D.C., 1995.

CHAPTER 9

MACHINE GUARDING AND WORKER TRAINING

... most managers know the importance of safety training, but it is not well known that specific requirements are detailed in OSHA, DOT, and EPA regulations. Under OSHA regulations, for example, it is stated or implied that the employer is responsible for providing training and knowledge to the worker. Moreover, employees are to be apprised of all hazards to which they are exposed, along with relevant symptoms, appropriate emergency treatment, and proper conditions and precautions of safe use or exposure.

... employers must comply with these standards and must require workers to comply. More than 100 OSHA ... safety and health regulations contain training requirements.

It is interesting to note that although OSHA requires training, it does not always specify what is required of the employer or entity that provides the training. However, information and instruction on safety and health issues in the workplace are foundational to building a viable organizational safety program. Workers cannot be expected to perform their assigned tasks safely unless they are aware of the hazards or the potential hazards involved with each job assignment.

— F. R. Spellman, p. 125, 1996.

9.1 INTRODUCTION

In October 1992, OSHA published an important lengthy document titled *Managing Worker Safety and Health.* This document suggests minimal requirements for an effective safety and health program. The four major aspects of its program approach are:

- Management Commitment and Employee Participation
- Worksite Analysis of Hazards
- Hazard Prevention and Control
- Safety and Health Training.

OSHA feels these four key elements are critical to developing an effective safety and health program (based on our experience, we agree). However, the last element on the list concerns us in this chapter: Safety and Health Training. Why? Though it is listed last in sequence above, we take the view (based on real-world experience) that training is the essence—the guts—of any effective safety program.

You don't need to be a neurosurgeon, an expert in quantum mechanics, or any other highly intelligent, highly educated individual to understand the importance of training. In the workplace or anywhere else in the real world, we simply cannot expect anyone to perform any job correctly and safely unless the person performing the function is capable, competent, qualified—and trained. Skills do not simply appear. For example, even a new employee with good experience does not know the requirements of a new facility. Directing a new employee to operate a complicated industrial machine, or to operate a bulldozer without first ensuring the would-be driver is capable, competent, qualified, or trained to properly operate the machine or the bulldozer is irresponsibile.

Irresponsibile? Why? What other term could we use to better describe such a situation? Certainly we're familiar with plenty of other terms we could use: negligent, criminal, uncaring, heartless, risky, dangerous, drifty, costly (sometimes very costly), idiotic—but to us, irresponsible seems to fit better—because it is one word that is all-encompassing—with little argument as to its exact meaning. Putting any person to work on any assignment without first training the worker to perform the job correctly and safely is irresponsible—and against the law; OSHA law.

We stress the need to ensure that the "worker" is properly trained. We also point out that safety training must be accomplished for **all** levels of employees (e.g., worker, supervisor, middle management, top management, owner) so that everyone knows the nature and extent of the hazards they work with, and their roles and responsibilities for establishing and maintaining safe work conditions. When we consider the most common cause of unsafe acts is lack of knowledge or information, we see that safety training is obviously not a novelty, but rather a necessity.

9.2 MACHINE SAFEGUARDING AND TRAINING

OSHA points out that even the most elaborate safeguarding system cannot offer effective protection—unless the worker knows how to use it and why. Specific and detailed training is therefore a crucial part of any effort to provide safeguarding against machine-related hazards. Thorough operator training should involve instruction or hands-on training, including:

- a description and identification of the hazards associated with particular machines
- the safeguards themselves, how they provide protection, and the hazards for which they are intended
- how to use the safeguards and why
- how and under what circumstances safeguards can be removed, and by whom (in most cases, repair or maintenance personnel only)
- what to do (e.g., contact the supervisor) if a safeguard is damaged, missing, or not capable of providing adequate protection.

This kind of safety training is necessary for new operators and maintenance or setup personnel, when any new or altered safeguards are put in service, or when workers are assigned to a new machine or operation.

> *Note: Again, employees should receive training on the machine and on the purpose of the safeguard before assignment to operate the machine.*

9.3 MACHINE GUARDING AND ADDITIONAL TRAINING

Training designed to instruct a worker on how to properly and safely operate a machine is only a part of the usual safety training. For example, certain machine operators must also be trained on protective clothing and personal protective equipment (PPE).

Engineering controls that eliminate the hazard at the source and do not rely on the worker's behavior for their effectiveness offer the best and most reliable means of safeguarding machinery. However, using engineering controls as the sole means of worker protection is not always possible; they may not be feasible or capable of completely protecting the employee. Under these circumstances, operators must wear protective clothing or PPE.

If protective clothing and/or PPE is to be effective, it must always be:

- appropriate for the particular hazards
- maintained in good condition
- properly stored when not in use, to prevent damage or loss
- kept clean, fully functional, and sanitary.

Protective clothing is, of course, available for different parts of the body. Hard hats can protect the head from falling objects, and caps and hairnets can help keep the worker's hair from being caught in machinery. If machine coolants could splash, or particles could fly into the operator's eyes or face, face shields, safety goggles, glasses, or similar kinds of eye protection might be necessary. Hearing protection may be needed when workers operate noisy machines. To guard the trunk of the body from cuts or impacts from heavy or rough-edged stock, protective coveralls, jackets, vests, aprons, and full-body suits are available. Workers can protect their hands and arms from the same kinds of injury with special sleeves and gloves. Safety shoes and boots or other acceptable foot guards can shield the feet against injury in case the worker needs to handle heavy stock which might drop.

> *Note that protective clothing and equipment can create hazards. A protective glove which can become caught between rotating parts, or a respirator facepiece which hinders the wearer's vision, for example, require alertness and continued attentiveness whenever they are used.*

Parts of the worker's ordinary clothing may present additional safety hazards. For example, loose-fitting shirts might possibly become entangled in rotating spindles or other kinds of moving machinery. Jewelry—bracelets and rings, for example—can catch on machine parts or stock and lead to serious injury by pulling a hand into the danger area, or catching on a protuberance and literally pulling off a finger. Long hair should be pulled back and secured. Workers need to dress according to the work they are performing, and eliminate these personal risks before beginning work.

Summary

Many of the safe work practices for operating machinery sound like good, solid common sense, when you think about them. But they are more than that—they are OSHA enforced, and that means compliance. We discuss OSHA's requirements for compliance in Chapter 10.

Cited Reference

Spellman, F. R., *Safe Work Practices for Wastewater Treatment Plants*. Lancaster, PA: Technomic Publishing Company, 1996.

CHAPTER 10

MACHINE GUARDING AND OSHA COMPLIANCE

OSHA is authorized under the OSH Act to conduct workplace inspections. Every establishment covered by the Act is subject to inspection by OSHA compliance officers.

Under the Act, "upon presenting appropriate credentials to the owner, operator or agent in charge," an OSHA compliance officer is authorized to:

Enter without delay and at reasonable times any factory, plant, establishment, construction site or other areas, workplace, or environment where work is performed by an employee of an employer; and to inspect and investigate during regular working hours, and at other reasonable times, and within reasonable limits and in a reasonable manner, any such place of employment and all pertinent conditions, structures, machines, apparatus, devices, equipment and materials therein, and to question privately any such employer, owner, operator, agent or employee.

With very few exceptions, inspections are conducted without advance notice.[9]

---

[9] *Adapted from F. R. Spellman,* Surviving an OSHA Audit. *Lancaster, PA: Technomic Publishing Company, pp. 1-3, 1998.*

10.1 INTRODUCTION

OSHA has the legal authority to enter most U.S. workplaces and to audit the site. Because of the OSH Act, OSHA compliance is of concern to many U.S. employers. Anyone who was not concerned until they were audited will tell you that a lack of concern with OSHA requirements can lead to very costly citations and possibly even to imprisonment.

Obviously, the employer is and should be concerned with penalties dispensed for non-compliance with OSHA regulations. But, actually, none of us should lose track of the number one motive for compliance with OSHA Standards—to ensure worker safety. A workplace in compliance with OSHA Standards usually has little to fear from OSHA—or from costly legal, medical, workers' compensation benefits, and/or other associated problems and costs. OSHA compliance thus gives employers double benefit, though that value is hard to determine until problems arise. Under OSHA compliance, the employer can prove his position.

On an audit, OSHA auditors make sure of compliance in the worksite, and they know what they are looking for. Machinery (in particular machine safeguarding) is a red flag to OSHA auditors. OSHA is well aware of the statistics related to workplace injuries (and worse, fatalities) related to the operation of unsafe machinery. Under an audit, employers can count on having machinery within the workplace closely scrutinized by the OSHA auditor.

In this chapter, we present a checklist approach—one that if properly used will help to ensure employers' machine safeguarding practices provide a safer workplace for employees and go a long way toward ensuring compliance with OSHA requirements.

The checklist approach is a valuable self-audit technique, one we advocate for all OSHA standards. However, self-audits do little good unless serious attention is paid to the checklist questions. Full answers are essential—and so is action on the areas that come up lacking.

10.2 REQUIREMENTS FOR ALL SAFEGUARDS

Answers to the following questions should help users determine the safeguarding needs of individual workplaces, by drawing attention to hazardous conditions or practices requiring correction for worker safety and OSHA compliance.

MACHINERY SAFEGUARDS CHECKLIST

	YES	NO
1. Do the safeguards provided meet the minimum requirements?	___	___
2. Do the safeguards prevent workers' hands, arms, and other body parts from making contact with dangerous moving parts?	___	___
3. Are the safeguards firmly secured and not easy to remove?	___	___
4. Do the safeguards ensure that no object will fall into the moving parts?	___	___
5. Do the safeguards permit safe, comfortable, and relatively easy operation of the machine?	___	___
6. Can the machine be oiled without removing the safeguard?	___	___
7. Is there a system for shutting down the machinery before safeguards are removed?	___	___
8. Can the existing safeguards be improved?	___	___

Mechanical Hazards

The point of operation:	YES	NO
1. Is a point-of-operation safeguard provided for the machine?	___	___
2. Does it keep the operator's hands, fingers and body out of the danger area?	___	___
3. Is there evidence that the safeguards have been tampered with or removed?	___	___
4. Could you suggest a more practical, effective safeguard?	___	___
5. Could changes be made on the machine to eliminate the point-of-operation hazard entirely?	___	___

Power transmission apparatus:	YES	NO
1. Are any unguarded gears, sprockets, pulleys, or flywheels present on the apparatus?	___	___
2. Are any belts or chain drives exposed?	___	___
3. Are any set screws, key ways, collars, etc. exposed?	___	___

	YES	NO
4. Are starting and stopping controls within easy reach of the operator?	___	___
5. Are separate controls provided for machines with more than one operator?	___	___

Other moving parts:	YES	NO
1. Are safeguards provided for all hazardous moving parts of the machine, including auxiliary parts?	___	___

Nonmechanical Hazards	YES	NO
1. Have appropriate measures been taken to safeguard workers against noise hazards?	___	___
2. Have special guards, enclosures, or personal protective equipment been provided, where necessary, to protect workers from exposure to harmful substances used in machine operation?	___	___

Electric Hazards	YES	NO
1. Is the machine installed in accordance with National Fire Protection Association and National Electrical Code requirements?	___	___
2. Are any conduit fittings loose?	___	___
3. Is the machine properly grounded?	___	___
4. Is the power supply correctly fused and protected?	___	___
5. Do workers ever occasionally receive minor shocks while operating any of the machines?	___	___

Training	YES	NO
1. Do operators and maintenance workers have the necessary training in how to use the safeguards, and why they should?	___	___
2. Have operators and maintenance workers been trained in where the safeguards are located, how they provide protection, and what hazards they protect against?	___	___
3. Have operators and maintenance workers been trained in how and under what circumstances guards can be removed?	___	___
4. Have workers been trained in the procedures to follow if they notice damaged, missing, or inadequate guards?	___	___

Protective Equipment and Proper Clothing	YES	NO
1. Is protective equipment required?	___	___
2. Is required protective equipment appropriate for the job, in good condition, kept clean and sanitary, and stored carefully when not in use?	___	___
3. Is the operator dressed safely for the job (i.e., no loose-fitting clothing or jewelry)?	___	___

Machinery Maintenance and Repair	YES	NO
1. Have maintenance workers received up-to-date instructions on the machines they service?	___	___
2. Do maintenance workers lock out the machine from its power sources before beginning repairs?	___	___
3. Where several maintenance persons work on the same machine, are multiple lockout devices used?	___	___
4. Do maintenance persons use appropriate and safe equipment in their repair work?	___	___
5. Is the maintenance equipment itself properly guarded?	___	___
6. Are procedures for lockout/tagout in place before workers attempt their tasks?	___	___
7. Are maintenance and servicing workers trained in the requirements of 29 CFR 1910.147, lockout/tagout hazard?	___	___

Summary

Checklists provide a starting point of examination for industrial facilities. They allow employers to examine existent machinery with thoughtful attention—allowing a different point of view and way of looking at familiar equipment and machinery, and a new perspective on how they are used. Familiarity and the thought that "we've always done it this way" can get in the way of seeing what should be done to provide safe working conditions to employees.

AFTERWORD

The rules to follow in safeguarding workers from accidental contact with machinery are straightforward. While a certain amount of ingenuity and creative thought may have to go into providing adequate safeguards for individual situations and machines, most situations can be solved simply. A complete hazard analysis reveals your facility's strengths and weaknesses, and directs your compliance efforts.

When you are examining machinery for hazard analysis, keep in mind these two essentials. Watch closely how your operators work, and listen to what they say about working on the machine. Not only will they provide you with the best information on how the machine and operation function together, any safeguard you install is worthless unless that worker buys into its use wholeheartedly. The best insurance you have for worker buy-in is worker participation. Change is difficult for anyone to accept, and changes that affect a worker's comfort with job tasks are prone to deliberate but quiet non-application, unless they meet the worker's needs. Human nature being what it is, they usually won't thank you for protecting them against their will, except under special circumstances.

The other way to combat worker non-compliance, of course, is thorough, complete and effective worker training, refresher training, and re-training. While employers complain that such training is time consuming and costly, the direct benefits are invaluable. Unfortunately, we cannot know the number of accidents that effective safe work practices prevent; we only see the accidents that occur.

The lessons learned from accident or injury are expensive ones, paid out in pain, and blood. We prefer less expensive lessons—the lessons effective machine safeguarding, training, hazard analysis, and self-auditing provide. Don't you?

APPENDIX A

Listing of Specific ANSI Safety Standards

ANSI B11.1-1982	Mechanical Power Presses
ANSI B11.2-1982	Hydraulic Power Presses
ANSI B11.3-1982	Power Press Brakes
ANSI B11.4-1983	Shears
ANSI B11.5-1988	Iron Workers
ANSI B11.6-1984	Lathes
ANSI B11.7-1985	Cold Headers and Cold Formers
ANSI B11.8-1983	Drilling, Milling, and Boring Machines
ANSI B11.9-1975	Grounding Machines
ANSI B11.10-1983	Metal Sawing Machines
ANSI B11.11-1985	Gear Cutting Machines
ANSI B11.12-1983	Roll Forming and Roll Bending Machines
ANSI B11.13-1983	Single- and Multiple-Spindle Automatic Screw/Bar and Chucking Machines
ANSI B11.14-1983	Coil Slitting Machines/Equipment
ANSI B11.15-1984	Pipe, Tube, and Shape Bending Machines
ANSI B11.17-1982	Horizontal Hydraulic Extrusion Presses
ANSI B11.18-1985	Machinery and Machine Systems for the Processing of Coiled Strip, Sheet, and Plate
ANSI B11.19-1990	Machine Tools, Safeguarding
ANSI B11.20-1991	Manufacturing Systems/Cells
(ANSI B15.1-1994/6)	Power Transmission Apparatus
ANSI B19.1-1990	Air Compressor Systems
ANSI B19.3-1986/90	Compressors for Process Industries
ANSI B20.1-1990	Conveyors and Related Equipment
ANSI B24.1-1985	Forging Machinery

ANSI B28.6-1983	Rubber Machinery, Hose
ANSI B28.7-1983	Rubber Machinery, Hose
ANSI B28.8-1983	Rubber Machinery, Hose
ANSI B28.9-1983	Rubber Machinery, Hose
ANSI B28.10-1986	Rubber Machinery, Endless Belt
ANSI B30.16-1987	Overhead Hoists
ANSI B151.1-1990	Plastics Injection Molding Machinery, Horizontal
ANSI B151.2-1982/88	Plastics Machinery, Film Casting
ANSI B151.3-1982/88	Plastics Machinery, Screen Changers
ANSI B151.4-1982/88	Plastics Machinery, Blown Film Takeoff and Auxiliary Equipment
ANSI B151.5-1982/88	Plastics Machinery, Film and Sheet Winding
ANSI B151.6-1982/88	Plastics Machinery, Slit Tape and Monofilament Postextrusion Equipment
ANSI B151.7-1982/88	Plastics and Rubber Extrusion Machinery
ANSI B151.11-1982	Plastics Machinery, Granulators, Pelletizers, and Dicers
ANSI B151.15-1985	Plastics Machinery, Extrusion Blow Molding
ANSI B151.21-1986	Plastics Machinery, Injection Blow Molding
ANSI B151.25-1998	Plastics Machinery, Injection Molding
ANSI B152.2-1982	Permanent-Mold Casting Machines (other than gray iron)
ANSI B153.1-1990	Automotive Lifts
ANSI B155.1-1986	Packaging Machinery

APPENDIX B

Suggested Reading

Accident Prevention Manual for Industrial Operations, 9th ed. Chicago: National Safety Council, 1988.

Alphabetical Index of Industrial Safety Data Sheets. Chicago: National Safety Council, 1988.

Cabinetmaking and Millwork. Feirer, John; Peoria, IL: Glenco Publishing Company.

Dictionary of Terms Used in the Safety Profession. Tarrants, W. E., ed.; Park Ridge, IL: American Society of Safety Engineers, 1988.

Disc Grinding-Safe Rules and Methods. Cleveland: Grinding Wheel Institute.

Electrical Standard for Metalworking Machine Tools. NFPA 79; Quincy, MA: National Fire Protection Association, 1987.

Fundamentals of Industrial Hygiene. Olishifski, J. B., and McElroy, F. E., eds.; Chicago: National Safety Council, 1988.

General Industry Safety and Health Standards. Title 29 Code of Federal Regulations 1910.212-1910.222.

Guards Illustrated. Chicago: National Safety Council, 1987.

Industrial Ventilation, A Manual of Recommended Practice, 20th ed.; Lansing, MI: American Conference of Governmental Industrial Hygienists, Edward Brothers, 1988.

Machine Tool Practice, White, W. T., and Kibbe, R. R., New York: Prentice Hall, 1991.

Power Press Safety Manual, Chicago: National Safety Council, 1989.

Safeguarding of Machinery, BS 5304. London: British Standards Institute, 1975.

The Use, Care, and Protection of Abrasive Wheels, ANSI B7.1-1988. New York: American National Standards Institute.

Safety Recommendations for Grinding Wheel Operation. Cleveland: Grinding Wheel Institute.

Safety Requirements for the Construction, Care and Use of Cold Headers and Cold Formers, ANSI B11.7-1985, New York: American National Standards Institute, 1985.

Safety Requirements for the Construction, Care and Use of Drilling, Milling, and Boring Machines, ANSI B11.8-1985. New York: American National Standards Institute, 1985.

Safety Requirements for the Construction, Care, and Use of Horizontal Injection Molding Machines, ANSI B151.1-1990. New York: American National Standards Institute, 1990.

Safety Requirements for the Construction, Care, and Use of Lathes, ANSI B11.6-1984. New York: American National Standards Institute, 1984.

Safety Requirements for the Construction, Care, and Use of Iron Workers, ANSI B11.5-1975. New York: American National Standards Institute, 1975.

Safety Requirements for the Construction, Care, and Use of Mechanical Power Presses, ANSI B11.1-1988. New York: American National Standards Institute, 1988.

Safety Requirements for the Construction, Care and Use of Metal Sawing Machines, ANSI B11.10-1990. New York: American National Standards Institute, 1990.

Safety Requirements for the Construction, Care, and Use of Packaging and Packaging-Related Converting Machinery, ANSI B155.1-1986. New York: American National Standards Institute, 1986.

Safety Requirements for the Construction, Care, and Use of Power Press Brakes, ANSI B11.3-1982. New York: American National Standards Institute, 1982.

Safety Requirements for the Construction, Care, and Use of Fasteners, Rivet Setting Equipment, ANSI B154.1-1984. New York: American National Standards Institute, 1984.

Safety Requirements for the Construction, Care, and Use of Shears, ANSI B11.4-1983. New York: American National Standards Institute, 1983.

Safety Requirements for the Construction, Care, and Use of Single and Multiple-Spindle Automatic Screw/Bar and Chucking Machines, ANSI B11.13-1983. New York: American National Standards Institute, 1983.

Safety Requirements for Woodworking Machinery, ANSI 01.1-1975. New York: American National Standards Institute, 1975.

Safety Specifications for Mills and Calendars in the Rubber and Plastics Industries, ANSI B28.1-1967. New York: American National Standards Institute, 1967.

Safety Standards for the Mechanical Power Transmission Apparatus, ANSI B15.1-1984. New York: The American Society of Mechanical Engineers, 1984.

Standard Handbook for Mechanical Engineers, 9th ed., Baumeister, T., ed., New York: McGraw-Hill, 1987.

www.ingramcontent.com/pod-product-compliance
Ingram Content Group UK Ltd.
Pitfield, Milton Keynes, MK11 3LW, UK
UKHW051418150926
475561UK00005B/223

9 780865 876620